ISO 14001
EMS Implementation
Handbook

ISO 14001
EMS Implementation
Handbook

Hewitt Roberts &

Gary Robinson

OXFORD BOSTON JOHANNESBURG MELBOURNE NEW DELHI SINGAPORE

Butterworth-Heinemann Ltd
Linacre House, Jordan Hill, Oxford OX2 8DP
225 Wildwood Avenue, Woburn, MA 01801-2041
A division of Reed Educational and Professional Publishing Ltd

℞ A member of the Reed Elsevier plc group

First published 1998

British Library Cataloguing in Publication Data
A catalogue record for this book is available from the British Library

Library of Congress Cataloguing in Publication Data
A catalogue record for this book is available from the Library of Congress

ISBN 0 7506 4020 0

Typeset by Butford Technical Publishing, Bodenham, Hereford
Printed and bound in Great Britain by
Biddles Limited, Guildford and King's Lynn

FOR EVERY TITLE THAT WE PUBLISH, BUTTERWORTH-HEINEMANN
WILL PAY FOR BTCV TO PLANT AND CARE FOR A TREE.

Printed on acid free paper

Contents

Preface

Who is this handbook for?

This handbook is for anyone interested in knowing, or needing to know, more about the implementation of corporate environmental management systems (EMS) compliant to ISO 14001 or EMAS. As corporate environmental management is a relatively new subject, it may still be difficult for you to assess whether you need to know!

Ask yourself the following questions:

- Are your customers expressing interest in the environmental implications of your products or services?

- Are your suppliers or clients becoming more concerned about the environmental implications of your products or services?

- Is it possible that your company or organization does not comply with all the environmental legislation and regulations that it should?

- Is environmental legislation regulating your industry sector increasing or becoming more stringent?

- Are the costs of energy, waste, waste treatment, water use, water treatment, or air pollution abatement of concern to your organization?

- Are other companies in your industry developing environmental management systems?

- Do you have international customers or clients?

If your answer to any of these questions was 'yes', then it is likely that you need to know more about environmental management and it is likely that your organization will benefit from the implementation of an environmental management system.

This handbook will provide you with all the information you need to plan, implement and maintain an environmental management system that meets the requirements of ISO 14001 and EMAS and it will help you start working towards corporate environmental improvement.

What this handbook provides

Step-by-step information on the requirements of a certifiable ISO 14001 EMS

This handbook has been designed to carefully take you through the steps of developing an environmental management system certifiable to ISO 14001 and EMAS. The chapters in this handbook consecutively address each component of the ISO 14001 standard from the identification of environmental aspects and impacts through to the management review.

Each chapter provides:

- *Introductory* information about the system component discussed in the chapter
- Point form *recommendations* for successful implementation of specific EMS requirements
- Blank, sample and explanation *templates* for EMS components that require templates (where applicable)
- '*Are you ready*' check-lists for each component developed to assess if you are ready for ISO 14001 certification

Internet-based, multimedia EMS case studies and an interactive support centre

To help take you smoothly through the EMS process, this handbook is complemented by an Internet web site at http://www.entropy-international.com/handbook/. The web site provides additional support and eases the transition from theory to practice.

The Internet site provides:

- Four real-life, multimedia case study examples of EMS development, implementation and maintenance. Each of the case studies illustrates a different stage in the EMS process from the initial review to final certification. The case studies are:

* United Distillers, Scotland: The initial environmental review
* Volvo Penta, Sweden: Maintaining your EMS
* Nortel, N. Ireland: Maintaining your EMS
* Aspects International, UK: Certifying your EMS

● Downloadable software to assist EMS implementation and documentation.

● On-line discussion groups on EMS, ISO 14001 and each of the ISO 14001 system requirements.

● Hundreds of internet links to up-to-date information on a variety of key topics including:

* ISO 14001
* EMAS
* Environmental legislation databases and subject-based search directories
* Environmental databases and subject-based search directories
* Chemicals, toxins, hazardous and special substance databases and search directories
* Environmental current events and industry news
* Environmental magazines, journals, periodicals and publications
* Environmental reporting
* Environmental technology
* Pollution prevention and cleaner production
* Jobs in the environmental industry
* Links to many additional sites of interest

How to use this handbook

This handbook has been designed to be used as an EMS implementation tool. For best results in implementing your EMS, this handbook should be used in conjunction with an official copy of the EMS standard requirements you wish to follow and the EMS handbook web site.

For information on how you can obtain an official copy of ISO 14001 please consult the ISO member organization closest to you. Contact information for these organizations is provided in Annex 4.

All of the templates, check-lists and questions used within this handbook are also provided in standard electronic formats on the floppy disk attached to the inside cover of the last page. These templates can also be downloaded directly from the handbook web site.

You will notice that icons are used throughout this handbook to accent important points visually. These icons are:

This icon indicates that a procedure is needed for successful EMS implementation.

This icon indicates that documentation is needed for successful EMS implementation.

This icon indicates the provision of an example from the United Distillers case study.

Introduction

What is environmental management?

Environmental management must simply be management of an organization's or company's impact on the environment. It seems straightforward but the fact that people have different views of what the environment is can cause confusion. Consequently, environmental management will likely mean different things to different people and thus an essential prerequisite to understanding environmental management would be an understanding of what the environment is.

An ecologist would probably say that the environment is 'all external conditions and factors, living and non-living, chemical and energy, that affect an organism or other specified system during its lifetime'[1]. An English teacher would likely say that the environment is 'a noun referring to our surroundings or external conditions that provide the conditions for life and growth'[2]. A sociologist may claim that the environment is 'the cultural, aesthetic and all other factors which contribute to the quality of life'[3].

Environmental management, therefore, is not necessarily as simple as it sounds and varies depending on what the 'environment' is. In many respects, the definition of environment, like the environment itself, depends on you. While there may never be one universal definition that applies to all people in all places, the definition of 'environment' in standardized approaches to environmental management, such as ISO 14001, requires some degree of consensus. As the purpose of this handbook is

---

[1] Miller, G. Tyler Jr., *Living in the Environment*, Eighth edition (Wadsworth Publishing, Belmont, CA, 1994).

[2] *English Dictionary* (Wm. Collins & Sons, London, 1983).

[3] ibid.

to provide you with the necessary skills and tools to develop, implement and maintain an environmental management system (EMS) certifiable to ISO 14001 (or verifiable to EMAS), an understanding of what is meant by 'environment' in those terms is essential. In ISO 14001 'environment' is defined as the 'surroundings in which an organization operates, including air, water, land, natural resources, flora, fauna, humans, and their interrelation'[4].

Consequently, environmental management and its desired result – improved environmental performance – is the process of minimizing the environmental impacts of your organization by controlling the aspects of your operations that cause, or could cause, impacts to that environment.

Improved environmental performance, like improved financial or quality performance, is a result by design, not by chance. Like all management systems, an EMS organizes resources to achieve certain objectives by establishing the procedures and infrastructures that, if followed and maintained, will yield a desired result. An EMS is no different. Its resources, objectives, procedures and infrastructures simply focus on improved environmental performance by controlling and minimizing the environmental impact of your company or organization.

Environmental management is not a 'hit and miss' approach to greening your company. Nor is it about replacing all machines, products and processes that have any impact on the environment. It is more akin to the Japanese philosophy of 'kaizen'[5] which is the relentless pursuit of gradual and unending improvement; only in this case it is a documented and planned process to improve environmental performance.

While this book focuses on the development of an EMS for an entire company or site wishing to certify to ISO 14001, it is essential to remember that an EMS can be as large or small as you choose. It can be formal and certifiable or informal and not certifiable. It can be developed over months or years. It can cover an entire company, a site, a process or even a single project. The choice is yours and should reflect the needs of your company.

The development and maintenance of a certifiable EMS is not rocket science. It is essentially the organized, documented, systematic and perpetual application of common-sense solutions to meet the desired objective of improved environmental performance.

---

[4] *ISO 14001 Environmental Management Systems – Specifications with guidance for use*, International Organization for Standardization (ISO) (Geneva, 1996).

[5] Hoyle, David, *ISO 9000 Quality Systems Handbook*, Second edition (Butterworth-Heinemann, Oxford, 1997).

While it is irrefutable that improved environmental management will benefit your company and is an essential ingredient to the social, economic and environmental sustainability of our planet, whether you choose to implement a full and standardized EMS for certification or a less formal system, this choice should reflect the present and future needs of your company or organization.

Some issues to consider when making this decision are as follows:

- All companies should have some sort of an EMS, even if informal and not certifiable.

- All EMSs should start with an initial environmental review of the site for which the EMS is being developed.

- All EMSs should have, at very least: an environmental policy; objectives and targets; management programmes; a register of legislation and regulation; and designated responsibility for the EMS.

- You can have a fully functional, standardized EMS without certification.

- Your standardized EMS can be certified immediately upon implementation, some time in the future or, as mentioned above, not at all.

- The certification process will increase the time, financial and human resources required for EMS development. What type of certification you choose or whether you certify at all should reflect the benefits your company will receive from certification (more in Chapter 1).

However, just as it is possible to manufacture cement life preservers made in a factory that has a certified quality management system (QMS), it is possible to develop an EMS for a factory that manufactures nuclear weapons. Having an EMS does not mean (and should not imply) that your company is environmentally benign. An EMS is merely a system that if used properly will enable your company to continually improve its environmental performance.

Your decision to implement an EMS (certifiable or not) must take into account a number of factors that affect your company or organization. While having a certifiable EMS is not necessarily appropriate for all organizations, the decision to develop an EMS certifiable to ISO 14001 can only be made by you and your company. When making this decision, be sure to sift through the criticisms of certifiable EMSs with care. Do not be dissuaded by drawbacks that could more properly be seen as disguised protectionism for the status quo of inadequacy or upper management's fear of what the results of EMS development might indicate. However, it is essential to remember that it is not the system or standard itself that improves corporate environmental performance; it is the individuals that operate those systems that do.

Acknowledgements

Special thanks to Agneta for her undying support and technical edits. Many thanks to Ian Lambart of United Distillers & Vintners for making so much of this possible and of course to David, Zina, Desta, and Mark & Gill for helping and supporting us along the way.

Chapter 1

Environmental management systems

Objective of the chapter

The objective of this chapter is to explain what environmental management and environmental management systems (EMS) are. This chapter will provide you with the skills necessary to begin the EMS process in your company. After finishing this chapter, you should be able to answer the following questions:

✓ What is environmental management?

✓ What is an environmental management system?

✓ Why is environmental management important?

✓ What are environmental aspects and impacts?

✓ What is ISO 14001?

✓ What is EMAS?

✓ What are the similarities and differences between ISO 14001 and EMAS?

✓ What are the similarities and differences between ISO 14001 and ISO 9000?

✓ What are the benefits of an EMS?

✓ What is the history of ISO 14001?

✓ What is the certification process for an EMS?

✓ What level of environmental management is necessary to meet the requirements of ISO 14001 and EMAS?

What is an EMS?

An environmental management system is the system by which a company controls the activities, products and processes that cause, or could cause, environmental impacts and in doing so minimizes the environmental impacts of its operations. As you can see from Figure 1.1, this approach is based on the management of 'cause and effect', where your company's activities, products and processes are the causes or 'aspects' and their resulting effects, or potential effects, on the environment are 'impacts'. Impacts would be things like a change in the mean temperature of a stream receiving effluent, an increased rate of asthma sufferers in a local population as a result of flue gas emissions, or contaminated land as a result of landfill leachate. Aspects would be the things within your company's control that cause, directly or indirectly, those impacts.

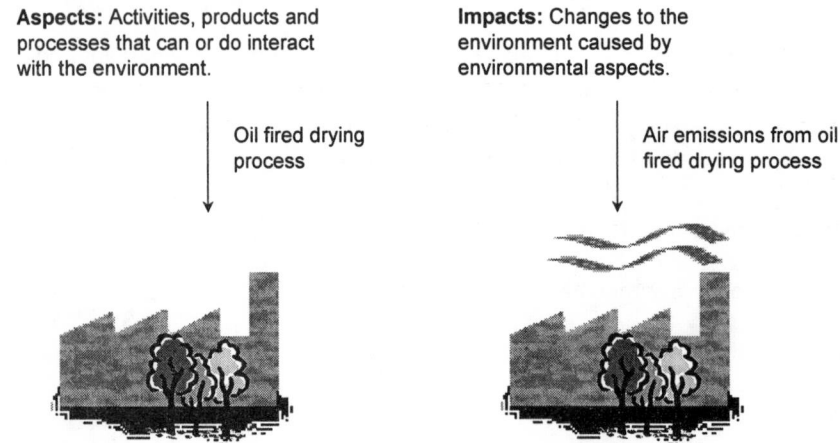

Aspects: Activities, products and processes that can or do interact with the environment.

Oil fired drying process

Impacts: Changes to the environment caused by environmental aspects.

Air emissions from oil fired drying process

Figure 1.1 *The definition of aspects and impacts – 'Cause and effect'*

Consequently, and as you can see in Figure 1.2, environmental management is essentially the tool that enables the control of aspects and thus minimizes and/or eliminates impacts.

Environmental management systems can be formal and standardized, such as ISO 14001 and EMAS (discussed later in this chapter), or they can be informal such as an internal waste minimization programme or the unwritten means and methods by which an organization manages its interaction with the environment.

Environmental management systems are very much related to quality management systems (QMSs). They are mechanisms that provide for a systematic and cyclical

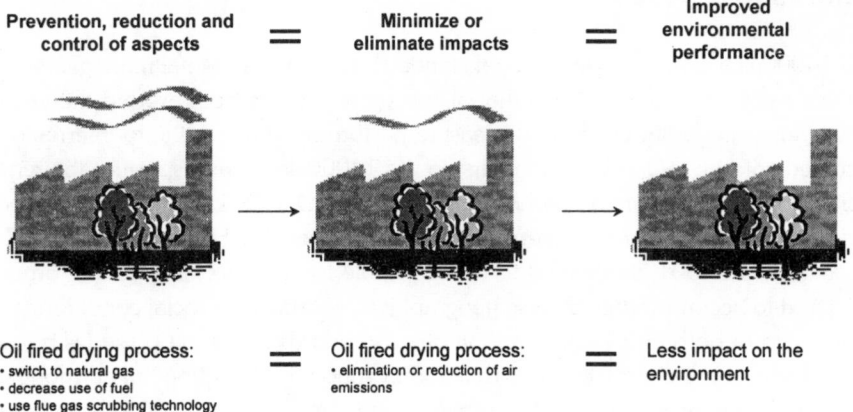

Figure 1.2 *Aspects, impacts and environmental performance – 'Prevention, reduction and control leads to improvement'*

process of continual improvement. As can be seen in Figure 1.3, the cycle itself begins with planning for a desired outcome (i.e. improved environmental performance), implementing that plan, checking to see if the plan is working and finally correcting and improving the plan based on observations from the checking process. Logically then, if the original outcome desired remains the same, a system of this nature will, by default, generate increments of progress that continually move toward the desired outcome.

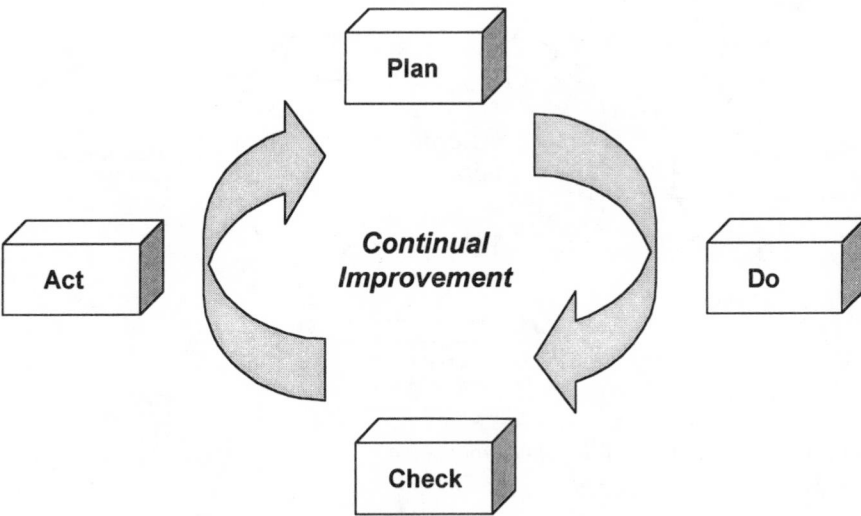

Figure 1.3 *Cycle of continual improvement*

What is ISO 14000?

ISO 14000 is a series of international standards for environmental management. It is the first such series of standards that allows organizations from around the world to pursue environmental efforts and measure performance according to internationally accepted criteria. ISO 14001 is the first in the 14000 series and specifies the requirements of an environmental management system. ISO 14001 is a voluntary standard and was developed by the International Organization for Standardization (ISO) in Geneva. ISO 14001 is intended to be applicable to 'all types and sizes of organizations and to accommodate diverse geographical, cultural and social conditions'[1]. The overall aim of both ISO 14001 and the other standards in the 14000 series is to support environmental protection and the prevention of pollution in harmony with socio-economic needs. ISO 14001 applies to any organization that wishes to improve and demonstrate its environmental performance to others through the presence of a certified environmental management system.

With the exception of requiring the *commitment* to continual improvement and *commitment* to comply with relevant legislation and regulation, ISO 14001 does not prescribe environmental performance requirements. The standard does not state the maximum allowable emission of nitrous oxide in flue gas or the maximum level of bacterial content in wastewater effluent. ISO 14001 specifies the requirements of the

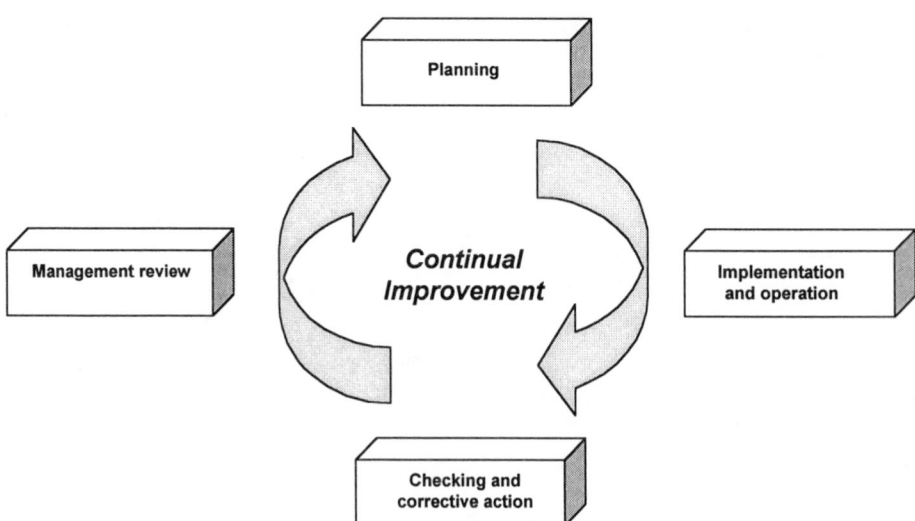

Figure 1.4 *Stages of ISO 14001 implementation*

---

[1] *ISO 14001 Environmental Management Systems – Specifications with guidance for use*, International Organization for Standardization (ISO) (Geneva, 1996).

management system itself, which, if maintained properly, will improve environmental performance by reducing impacts such as NOx emissions and bacterial effluents.

The requirements of an EMS certifiable to ISO 14001 and its correlation with the aforementioned cyclical approach to continual improvement (plan, do, check, correct) are reflected in Figure 1.4 above.

The ISO 14000 series includes the following standards and proposed standards:

Standard	Title
14001	Environmental Management Systems - Specifications with Guidance for Use
14002	Environmental Management Systems - Guidelines on Special Considerations Affecting Small and Medium Scale Enterprises
14004	Environmental Management Systems - General Guidelines on Principles, Systems and Supporting Techniques
14010	Guidelines for Environmental Auditing - General Principles of Environmental Auditing
14011	Guidelines for Environmental Auditing - Audit Procedures Part 1: Auditing of Environmental Management Systems
14012	Guidelines for Environmental Auditing - Qualification Criteria for Environmental Auditors
14013/15	Guidelines for Environmental Auditing - Audit Programmes, Reviews and Assessments
14020	Environmental Labels and Declarations - General Principles
14021	Environmental Labels and Declarations - Environmental Labelling - Self Declaration of Environmental Claims - Terms and Definitions
14022	Environmental Labels and Declarations - Environmental Claims - Self Declaration of Environmental Claims - Symbols
14023	Environmental Labelling - Self Declaration of Environmental Claims - Testing and Verification Methodologies
14024	Environmental Labels and Declarations - Environmental Labelling Type I - Guiding Principles and Procedures
14031	Environmental Performance Evaluation - Guidelines
14032	Technical Report Type III - Environmental Management - Environmental Performance Evaluation - Case Studies Illustrating the Use of ISO 14031
14040	Life Cycle Assessment - Principles and Framework
14041	Life Cycle Assessment - Life Cycle Inventory Analysis
14042	Life Cycle Assessment - Impact Assessment
14043	Life Cycle Assessment - Interpretation
14049	Technical Report Type III - Environmental Management - Life Cycle Assessment - Examples for the Application of ISO 14041
14050	Environmental Management Terms and Definition
14061	Technical Report Type III - Guidance to Assist Forestry Organisations in the Use of ISO 14001 and ISO 14004

What is EMAS?

Although ISO 14001 is the only international standard for an EMS, there are other standards that prescribe requirements for a functional EMS. One of the first and more recognized standards is EMAS, a European Union Regulation regarding environmental management systems. This is Council Regulation No. 1836/93, of 29 June 1993, which allows voluntary participation by companies in industrial sectors in a Community Eco-Management and Audit Scheme (EMAS). EMAS requires EU member states to set up supporting administrative structures for the scheme and allows companies to participate on a voluntary basis.

The overall aim of EMAS was to meet the European Community's obligation to develop 'policy and action in relation to the environment and sustainable development'[2] as stipulated in the Treaty of the European Union signed at Maastricht in 1992. In conjunction with preceding Council resolutions prescribing the roles and responsibilities of companies both to reinforce the economy and protect the environment, EMAS recognizes that industry has its own responsibility to manage the environmental impact of its activities and therefore should:

- Adopt a proactive approach in this field.

- Prevent, reduce and as far as possible eliminate pollution, particularly at the source.

- Ensure sound management of resources.

- Use clean or cleaner technologies.

EMAS prescribes that this responsibility calls for companies to establish and implement effective environmental management systems including, among other things, an environmental policy, objectives, programmes and the provision of information on environmental performance to the public (the environmental statement) all aimed at reasonable continuous improvement of environmental performance.

Comparison between ISO 14001 and EMAS

The EMAS Regulation states: 'To avoid unjustified burdens on companies and ensure consistency between the Community scheme and national, European and international standards for environmental management systems and audits, those standards recognized by the Commission shall be considered as meeting the corresponding requirements' of EMAS[3]. However, at the time of writing, ISO 14001 was not officially considered as meeting those requirements. This was due mostly to the fact that ISO 14001 is a very recent standard and that there are issues of substance, clarification and scope that had to be considered before EMAS registration could be granted to organizations certified to ISO 14001.

---

[2] Council Regulation (EEC) No 1836/93 of 29 June, 1993 allowing voluntary participation by companies in the industrial sector in a Community eco-management and audit scheme – *Official Journal of the European Communities* (10 July, 1993), OJ 10.7.93 L 168.

[3] ibid.

Recently, however, the European Commission's Regulatory Committee published a 'Bridging Document' that details the differences between the two systems. In this document it was formally stated that: 'EMAS Registration can be obtained by companies with a certified ISO 14001 system who demonstrate, to an accredited EMAS verifier, that their EMS meets the requirements of the EMAS Regulation'[4].

Some of the more salient differences are as follows:

Differences between ISO 14001 and EMAS

- ISO 14001 is a globally applicable standard whereas EMAS is a Regulation for the participation of companies in EU member states.

- EMAS specifically requires the completion of an initial environmental review before implementing the EMS whereas ISO 14001 only suggests that carrying out such an exercise is useful for developing an EMS and that significant environmental impacts and aspects should be identified (more on the initial environmental review in Chapter 2).

- EMAS requires the preparation of an environmental statement to be made publicly available. This statement must be verified externally to ensure reliability of the information. There is no requirement for an environmental statement in ISO 14001. (N.B. Environmental statement should *not* be confused with the environmental policy or other mandatory reporting requirements.)

- Any type of organization may certify to ISO 14001 whereas EMAS is limited to industries within a specified scope. Originally, EMAS was applicable only to the industrial sector (mining and quarrying; manufacturing; electricity, gas and water supply; and solid and liquid waste) but has since extended its scope considerably and now includes service industries and local governments.

- ISO 14001 can be applied to a whole company, a site or even specific activities within a company or site whereas EMAS is applicable only to a 'site'.

- EMAS states that a company must 'provide for compliance with all relevant requirements regarding the environment' while ISO 14001 only states that there must be a 'commitment to comply' with relevant environmental legislation and regulations.

- EMAS states that the audit of environmental performance and management system is performed, or the audit cycle completed, at least every three years while there is no audit frequency specified in ISO 14001.

- EMAS states that your environmental policy must include a commitment to the continuous improvement of environmental performance with a view to reducing impacts to levels not exceeding those corresponding to economically viable application of best available technology (EVABAT) while ISO 14001 states that the EMS should encourage the use of best available technology where appropriate and economically viable.

---

[4] CEN. Revised text for the *Bridging Document Between EMAS and EN ISO 14001, 1410, 14011 and 14012* – unpublished (1996).

Comparison between ISO 14000 and ISO 9000

ISO 14000 is a series of international standards for environmental management systems while ISO 9000 is a series of international standards for quality management systems. ISO 9000 preceded the 14000 series of standards and was similarly published by the International Organization for Standardization (ISO) in Geneva.

ISO 9000 was developed to help companies meet customer requirements through a systematic control of the production process while aiming to improve continually. ISO 14001 is meant as a tool to help companies continually improve their environmental performance by controlling and minimizing identified environmental impacts of their operations.

ISO 9000 and 14001 are very similar in structure and have a number of common elements such as policies, training, operational control, document control, audits, nonconformance, correction and prevention. Both ISO 9000 and 14001 should be seen as part of an organization's overall management structure and integration of all management is encouraged where possible.

ISO 9000 and 14001 are so similar, in fact, that ISO itself, at the time of writing, is addressing integration of the two standards. Consequently, it is fair to say that if your company has already established an ISO 9000 system, you are well on your way to the development of an ISO 14001 system for environmental management. While there are a number of similarities between ISO 9000 and 14001, one should not be seen as paramount to or to obviate the need for the other.

The following table outlines some of the major similarities and differences between ISO 9000 and ISO 14001.

Similarities between ISO 14001 and ISO 9000

- Both 9000 and 14001 provide specifications for a management system within an organization.

- Both 9000 and 14001 specify the need for a policy as a guiding management document.

- Both ISO 9000 and ISO 14001 specify the need for an established organizational structure.

- Both ISO 9000 and ISO 14001 specify the need for operational control.

- Both ISO 9000 and ISO 14001 specify the need for document control.

- Both ISO 9000 and ISO 14001 specify the need for corrective and preventative action.

- Both ISO 9000 and ISO 14001 specify the need for record keeping.

- Both ISO 9000 and ISO 14001 specify the need for training within the organization.

- Both ISO 9000 and ISO 14001 specify the need for system audits.

Differences between ISO 14001 and ISO 9000

- ISO 9000 addresses quality management whereas ISO 14001 addresses environmental management.

- ISO 9000 addresses customer requirements whereas ISO 14001 addresses environmental performance and the needs of a far wider range of stakeholders.

- ISO 14001 stipulates commitment to comply with relevant environmental legislation, regulations and industry codes of practice.

- ISO 14001 requires the identification of your organization's significant environmental aspects and impacts.

- ISO 14001 specifies the need for the setting of environmental objectives and targets.

- ISO 14001 specifies the need for emergency preparedness and response.

- ISO 14001 specifies the need for a publicly available environmental policy and a means for communicating internally and externally with respect to environmental aspects and impacts.

Why develop an EMS?

As an aspiring environmental manager or environmental manager-to-be, you will undoubtedly be confronted with the question: 'Why do we need an environmental management system?' While the answer may be obvious to you, it will most certainly not be obvious for many of your peers, colleagues or workplace superiors. Accepting that one of the most fundamental ingredients of a successful EMS is top management commitment, one of the first tools required for the task at hand is a clear understanding of the advantages of implementing an EMS.

The advantages of improved environmental management can be divided into two broad categories. The first category addresses the fact that improved environmental management is good for our planet and a fundamental requirement of global sustainability. The second category addresses the fact that improved environmental management could be seen as a future requirement of sustainable commerce and good for your business. One of these reasons will be more appealing to your boss or board of directors and it should not take too long to figure out which one.

Obviously, while saving the planet is meritorious and essential, the second reason is likely to improve your chances of convincing others to provide commitment and resources to the development of an EMS. However, before ascertaining what corporate advantage you may be able to gain by developing an EMS, it is important to at least consider the larger objective of corporate environmental management.

The benefits for our planet

Global economic activity is now valued at well over 20 trillion US dollars annually[5]. All of this economic activity either draws from the resources of the planet or emits to our global environment. Both economic and population growth are accelerating at exponential rates and rapidly approaching threshold limits. Respecting that present business patterns are fundamentally unsustainable, improved environmental management will serve at least to move our business patterns towards sustainability. To help understand the magnitude of this problem, consider the following facts:

---

[5] Hawken, Paul, *The Ecology of Commerce* (Harper Collins, New York, 1994).

- 83% of the commercial energy used in the world comes from non-renewable sources[6].

- Every day the world-wide economy burns an amount of energy that the planet required 10,000 days to create[7].

- 0.003% of Earth's supply of water is useable as fresh water[7].

- Between 65% and 70% of the world's fresh water is wasted[7].

- Today the world emits over 100 million metric tonnes of sulphur into the atmosphere annually. This leads to over 200 million metric tonnes of SO_2 emissions and is the major contributor to acid rain[8].

- Acid rain is believed to contribute to European crop losses of over $500 million annually[8].

- Accumulation of greenhouse gases will cause a global temperature rise of 1.5C – 4.5C over the next century[9].

- The world uses over 4 billion pounds of pesticides a year[7].

- 25 million people world-wide are poisoned annually by pesticides used in agriculture and farming[7].

- Human activities introduce over 100,000 chemicals and toxins into the Earth's environment annually[7].

- 56% of the world's tropical forests have already been destroyed by human activity[6].

- Approximately 340,000 square kilometres of tropical forests are destroyed or degraded every year[6].

- Up to 50% of the world's wetlands have been destroyed by human activity[6].

- 33% of the world's cropland is eroding faster than it forms[6].

- At least 40,000 children in less developed countries die every day from preventable diseases[6].

- At least 1.3 billion people world-wide are undernourished or underfed[6].

- Nearly 100 of Earth's species become extinct every day due to human activities[6].

'Quite simply, present business practices are destroying life on earth'[7].

[6] Miller, G. Tyler Jr., *Living in the Environment*, Eighth edition (Wadsworth Publishing, Belmont, CA, 1994).

[7] Hawken, Paul, *The Ecology of Commerce* (Harper Collins, New York, 1994).

[8] Bellini, James, *High Tech Holocaust* (Sierra Club, San Francisco, CA, 1986).

[9] Hileman, Bette, *Web of Interaction Makes it Difficult to Untangle Global Warming Data* (C&EN, Washington, 1992).

The benefits for your organization

It is commonly accepted that environmental issues are increasingly affecting both the long and short-term performance of organizations. Similarly, environmental issues affect both revenues and costs. Poorer environmental practices lead to higher manu-facturing and non-manufacturing costs; higher quantities of spoils and waste; increased cost of waste disposal; the expense of abatement technologies; environ-mental fines and mitigatory public relations campaigns; and higher insurance premiums. The list is growing and complex.

Some of the benefits that can be expected from improved environmental perform-ance are given below.

Cost savings

Organizations encouraging initiatives to improve overall environmental performance such as environmental management systems, cleaner technologies or waste mini-mization programmes have demonstrated the ability to generate considerable savings. The ISO 14001 implementation process will enable you to identify resource use and inefficiencies therein and provide for a framework to evaluate opportunities and possibilities for cost saving.

For example:

- Between 1975 and 1990, 3M corporation saved over $537 million by implementing environmental initiatives in its organization[10].

- Project Catalyst, a UK Department of Trade and Industry (DTI) demonstration proj-ect, identified potential savings of £8.9 million from 399 waste minimization measures in 14 large and small enterprises[11].

- The Aire and Calder project, a UK-based initiative sponsored by the BOC Foundation for the Environment, identified savings of £3.3 million a year for the 11 participating companies[12].

- According to the car manufacturer, Rover, six of its suppliers recently implemented environmental management systems, making cost savings of between £10,000 and £100,000[13].

---

[10] Hawken, Paul, *The Ecology of Commerce* (Harper Collins, New York, 1994).

[11] *Project Catalyst: Report to the Project Completion Event at Manchester Airport*, 27 June, 1994 (WS Atkins, March Consulting & Aspects International, Manchester, 1994).

[12] *Waste Minimization: A Route to Profit and Cleaner Production* (UK Centre for the Exploitation of Science and Technology, 1994).

[13] *Small Firms and the Environment* (Groundwork Trust, UK, 1995).

Clearly, improved environmental management will identify short and long-term savings and opportunities, and prepare a business to respond to future environmental pressures.

Increased efficiency

In addition, and closely related to cost savings, the implementation of an EMS also increases a firm's efficiency. Whether it is better use of raw material or improved product quality, an EMS provides an organization with an overview of its operations and enables process improvement and an increase in efficiency. Similarly, the development of an EMS will enable you to identify and correct other internal management problems where they may exist and provide for efficiency through operational integration with other management systems in your company.

Increased market opportunities

One of the fundamental reasons for the development of ISO 14001 was to reduce non-tariff trade barriers while simultaneously generating a commitment to environmental performance world-wide. Consequently, developing an internationally accepted environmental management system has clear international marketing benefits. An ISO 14001 EMS can not only preserve an organization's position in international markets but can serve as a ticket to new ones. An EMS demonstrates to customers that your company has made a commitment to environmental performance that they are growing to expect. Having a certified EMS may also serve as a leg up in winning procurement bids and contracts from international clients and governments that have similarly made a commitment to environmental performance. 'Effective environmental management is a key aspect of good business practice allowing companies to take advantage of marketing opportunities as well as controlling the environmental impacts of their operations'[14].

Increased ability to comply with environmental legislation and regulations

As you will see in Chapter 5, *Legislation and regulations*, one of the fundamental requirements of ISO 14001 is knowledge of and commitment to comply with environmental legislation and regulations relevant to your company. Consequently, a functional EMS is most definitely a step in the right direction to ensure that your company stays on the right side of the law. Furthermore, an EMS demonstrates to authorities and regulators that you are, at the very least, committed to compliance and will often improve relations with them.

---

[14] *EMAS – Positioning Your Business* (Business in the Environment and Coopers & Lybrand, London, 1995).

Meeting the demands of your clients

As the development of an EMS requires that you attempt to extend the responsibility of improved environmental performance to your suppliers, with the number of certified EMSs growing world-wide, there are similarly a growing number of companies beginning to feel 'business-to-business pressures' to demonstrate some form of corporate environmental management. Business-to-business pressure is simply when company A, usually a larger, certified and revered customer of company B, tells company B, in gentle but unmistakable terms, that unless they implement an EMS in X amount of time, they can consider their services no longer needed. Alleviating 'business-to-business pressures' by meeting the environmental demands of your customers is thus another clear benefit of EMS implementation.

Improved relations with stakeholders

In addition to the other more tangible benefits of implementing an EMS, an environmental management system also creates a number of 'soft' benefits. Of growing significance is the fact that implementing an EMS improves a company's relationship with its many stakeholders (neighbours, shareholders, customers, clients, bankers, insurers etc.)

The development of an EMS improves stakeholder relations both directly and indirectly. Directly, an EMS decreases a company's impact on the environment, pleasing neighbours and pressure groups. It reduces risk and liability, pleasing employees and insurers, and it increases profitability, which of course pleases the shareholders and the bank manager.

An EMS also improves stakeholder relations indirectly through certification of your EMS. In this case, regulators, authorities, clients and customers need not concern themselves with overly exhaustive inspections, assessments or investigations as legislative compliance (or at least an attempt to comply) and a stated desire for environmental improvement are prerequisites for certification. In short, the development of an EMS provides companies with an externally visible stamp of approval that demonstrates to their stakeholders that they are taking steps to manage their environmental impact.

Increased motivation, loyalty and commitment from, and communication with, employees

Another benefit associated with the implementation and maintenance of an EMS is that of increased employee motivation, productivity and loyalty. While an employee is certainly a stakeholder and thus enjoys the aforementioned stakeholder benefits, the EMS process uniquely affects them. This process forces a firm to assess a number of factors of great importance to any workforce. Worker health and safety, risk and emergency situations, education and training are all issues that must be considered when developing and maintaining an EMS.

This process, as stated in the standard, must involve all workers. It encourages participation, facilitates improved communication and is a co-operative effort for a unified purpose. This process humanizes and harmonizes. It bridges gaps among rank and through its reliance on participation is a vehicle for improved self worth, job satisfaction and productivity.

While the opportunities provided through the EMS process are not unlimited, they are wide and varied, direct and indirect, hard and soft, and while it is not necessary to expose them all, it is important to appreciate that they are possible and numerous. In short, 'every business must have some sort of management system in order to operate and survive, and the step to incorporating environmental management, even if only to a limited extent, will strengthen existing systems, cut costs and increasingly be a necessity for business survival'[15].

History of standardized EMS

While the exact origins of corporate environmental management are not certain, it is generally accepted that the ISO 14000 series of standards emerged as a result of both the Uruguay round of the GATT negotiations and the United Nations Rio Summit on the Environment held in 1992. One of the premises of GATT is the reduction of non-tariff barriers to trade and one of the main purposes of the Rio Summit was to develop global commitment to sustainability and the protection and enhancement of the environment.

Hence, with the growing application of ISO 9000 and the increasing development of national standards for environmental management in the 1980s, the International Organization for Standardization (ISO) recognized the need to assess the applicability of an international standard for environmental management. Consequently, the Strategic Advisory Group on the Environment (SAGE) was formed in 1991 to determine whether a standard for environmental management could:

- Promote a common approach to environmental management similar to that of ISO 9000 and quality management.

- Enhance an organization's ability to attain and measure improvements in environmental performance.

- Facilitate trade and remove trade barriers[16].

---

[15] *BS 7750? Heard of it?* (Environment Times, UK, 1995).

[16] *Quality Network, ISO 14000 Introduction*, www.quality.co.uk/quality/iso14000.htm (UK, 1996).

The Group's findings promoted the development of the ISO 14000 series of standards very much in line with the ISO 9000 standards for quality.

While it is arguable that environmental management has been an integral part of most, if not all, indigenous cultures of the world and that quality management has been around since the Egyptians built pyramids, formal and documented corporate management systems for quality developed as an element of industry after WWII. By 1979 the British Standard Institute (BSI) published the three-part quality series BS 5750. BS 5750 closely resembled earlier defence standards and by the mid-1980's most industrial countries had similar standards in place. As the international importance of a standardized approach to quality management grew, the ISO developed ISO 9000. ISO 9000 was first published in 1987 and bears significant resemblance to BS 5750 and earlier quality standards for munitions manufacturing in the defence industry[17].

Following closely behind the development of BS 5750 was BS 7750, the first formal, systematic and standardized approach to environmental management. BS 7750, like BS 5750 was published by the BSI and was a voluntary management standard. BS 7750 was first published in March of 1992 under the title *British Standard: Specifications for environmental management systems.*

As a basis for the EMS, BS 7750 required commitment to continuous improvement and compliance with relevant legislation as a key starting point. BS 7750 placed a large emphasis on the environmental policy in providing direction for the development and maintenance of the EMS. It is important to understand that BS 7750 did not establish specific requirements for environmental performance, beyond compliance with relevant legislation and regulations and a commitment to continual improvement but rather maintained that organizations develop environmental policies, objectives and targets, and procedures to control and minimize their identified significant environmental effects. Just as the international quality management standard (ISO 9000 series) was developed directly from BS 5750, ISO 14001 and EMAS have been developed directly from BS 7750.

Although it must be said that improved corporate environmental performance is paramount to social, economic and environmental sustainability, it is saddeningly paradoxical to realize that the basis for ISO 14001 is not centuries of knowledge about environmental sustainability but rather humankind's insatiable need for military hardware.

---

[17] Hoyle, David, *ISO 9000 Quality Systems Handbook*, Second edition (Butterworth-Heinemann, Oxford, 1997).

EMS certification

What is certification?

As discussed earlier, ISO 14001 prescribes the requirements for a system, not environmental performance itself. Similarly, certification is of the management system itself, not environmental performance. An audit is *not* conducted to ascertain whether your flue gas emissions are less than X part per million nitrous oxide or that your wastewater effluent contains less that Y milligrams of bacteria per litre. Consequently, the process of auditing the system for compliance to the standard entails checking to see that all of the necessary components of a functioning system are present and working properly.

As also discussed earlier, your company can have a complete and fully functional EMS as prescribed by ISO 14001 without being certified. As certification can add to the time and expense of EMS development, it is important for you to establish, in advance, whether certification is of net benefit to you. Although most companies that develop an EMS do in fact certify, there are cases where certification does not add immediate value. Certification is not always beneficial to small and medium sized companies. Certification is not always necessary for companies with one or two large clients with environmental demands who are satisfied that you have a functional EMS (second-party declaration). Whatever decision you make, it is important to remember that just as a driver's licence does not automatically make you a good driver, ISO 14001 certification does not automatically make your company environmentally benign or ensure that you will continually improve environmental performance. The system is only as good as the people who operate it.

While some companies prefer not to certify their EMS immediately, most in fact do and for a number of very good reasons, some of which are as follows:

- A certificate is proof of evaluation and acceptance by an independent, accredited and third-party professional.

- A certificate can be seen as an external stamp of approval of your EMS and your commitment to improved environmental performance.

- A certificate will be beneficial in winning international and governmental procurement contracts.

- A certificate can obviate lengthy legislative and regulatory compliance audits.

- A certificate serves as a visible symbol of your company's intentions with respect to the environment.

- The periodic assessments from your certifier will serve as a motive for continued maintenance, improvement and integrity of your EMS.

The ISO 14001 certification process

When you reach the point where your EMS is meeting, or very close to meeting, the specifications outlined in ISO 14001, you can do one of the following:

1 Self declare that your EMS conforms to the standard.

   Self declaration means that your company audits its own EMS against the specifications set out in the standard and 'declares' that it meets the requirements specified. As it does not involve independent auditors, this approach can be of limited value to external parties.

2 Seek second-party recognition that your EMS meets the standard requirements.

   Second-party recognition is when an organization other than your own, such as a client or supplier, declares that your EMS meets the requirements of the standard; i.e. Company A lets Company B audit their EMS to satisfy demands that Company A has a functional EMS. Second-party recognition may be valuable when a client or supplier requires your company to have an EMS but recognizes that a formal certificate is not necessary.

3 Seek formal third-party certification from an independent accredited certification body.

   Third-party recognition is when you pay an external, autonomous and independent accredited certification body to audit your EMS and officially declare that your EMS meets the requirements of ISO 14001. Third-party certification is most common and offers certifiable proof that your system conforms to the specifications outlined in ISO 14001.

While the exact process of formally certifying your EMS to ISO 14001 will depend on the accredited certification body that you work with, the following points summarize the major steps that you will need to take to receive third-party certification (Figure 1.5).

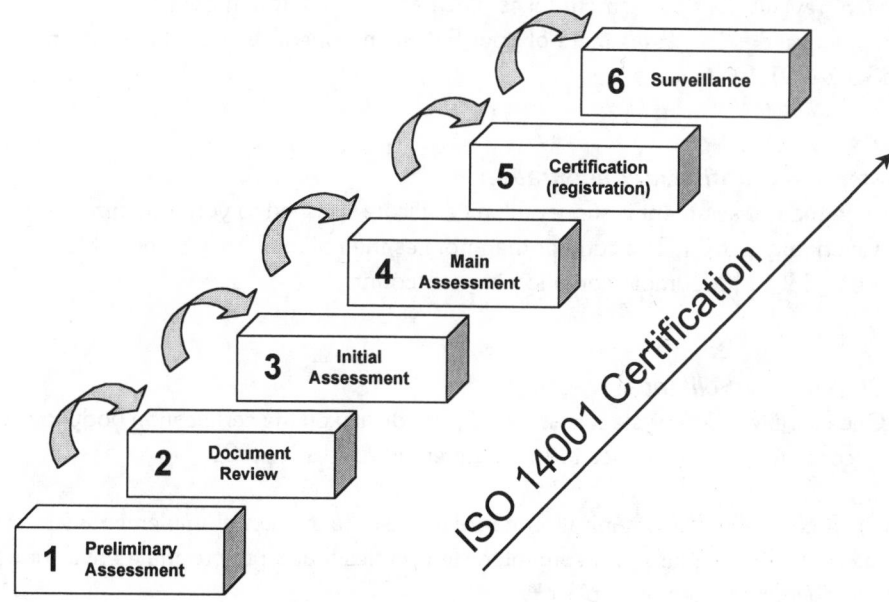

Figure 1.5 *The main steps towards ISO 14001 certification*

Step 1 – Preliminary assessment

Some certifiers offer the choice of a preliminary assessment or 'gap analysis' between your EMS and the main requirements of ISO 14001. This generally helps to identify problem areas before the main certification assessment.

Step 2 – Document review

An off-site audit of your EMS documentation ensures that the essential documents such as your environmental policy, objectives and targets, registers, procedures etc. are present and properly prepared.

Step 3 – Initial assessment

Having passed the document review and implemented any recommendations made, a site visit is conducted to ensure preparedness for the main assessment and allow the certifier to gain a better understanding of the EMS and those directly involved with it.

Step 4 – Main assessment

The main and most in-depth assessment of the EMS is conducted after allowing for changes to the system in light of any findings from the initial assessment and document review. This assessment takes place at your site (often over several days) and involves a detailed assessment of your EMS components against the requirements of ISO 14001.

Step 5 – Certification/registration

If the main assessment is successful, a certificate is issued to your organization by the accrediting certifier. The certifier then notifies the national body responsible for overseeing ISO 14001 implementation in your country.

Step 6 – Surveillance

Once certified, your system is assessed periodically by the certification body to ensure its continuing conformance to the requirements of ISO 14001.

Once certified your organization can demonstrate successful implementation of the international standard to assure interested parties that an appropriate environmental management system is in place[18].

The EMAS verification process

In many respects the process of certifying your EMS to the requirements of EMAS is essentially the same as the process of certifying an EMS to ISO 14001. However, there are two fundamental differences. One (typical) difference is the terminology used. In ISO language the process uses the terms *certification* and *registration* while in EMAS the terms *verification*, *validation* and *registration* are used. A second difference is that EMAS, unlike ISO 14001, requires the preparation of an environmental statement that becomes an important component of the process.

After the development and implementation of a functional EMS the steps to verification and registration are as follows (Figure 1.6).

Step 1 – Preparation of the environmental statement

A requirement of EMAS is that you must prepare an environmental statement upon completing your first EMS audit. The environmental statement must be written in clear and non-technical language and designed for the public.

---

[18] *ISO 14001 Environmental Management Systems – Specifications with guidance for use*, International Organization for Standardization (ISO) (Geneva, 1996).

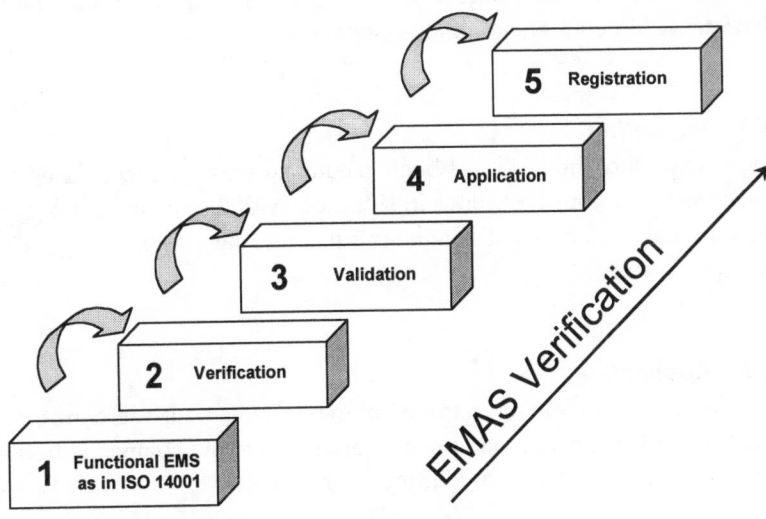

Figure 1.6 *The main steps towards EMAS verification*

Furthermore, the statement should:

- Provide information indicating that your environmental policy is sound.

- Provide information indicating your objectives and targets and your environmental management programmes are coherent.

- Describe the activities of your company at the site being certified.

- Include an assessment of the significant environmental issues of the site.

- Include a summary of the figures on pollution emissions, waste generation, consumption of raw materials, energy use, water use, noise and other significant environmental aspects of the site.

- Include an assessment of the factors regarding environmental performance of the site.

- Include your environmental policy.

- State a deadline for submission of the next environmental statement.

- State the name of your intended accredited verifier.

Step 2 – Verification

Having prepared your environmental statement it is then necessary to have an independent and accredited verifier audit the environmental policy, environmental programmes and environmental audit. Verifiers ensure that these components are present and meet the requirements of the EMAS Regulation and that they are accurately presented in your environmental statement.

Step 3 – Validation

Having verified that the EMS has been accurately presented in your environmental statement and that your EMS does in fact meet with the requirements of the EMAS Regulation, the environmental statement is then validated (signed) by the independent accredited verifier.

Step 4 – Application

Upon successful validation of your environmental statement and corresponding site, the validated environmental statement is forwarded to the competent body responsible for the EMAS Regulation in your member state.

Step 5 – Registration

The competent body then registers acceptance of the validated statement and informs you of this by providing you with a registration number.

You are then allowed to use (for the registered sites only) the EMAS logo and one of the four official statements of participation.

The statement that applies to companies and individual sites is: 'This site has an environmental management system and its environmental performance is reported on to the public in accordance with the Community Eco-Management and Audit Scheme.'[19]

When using the statement of participation:

- You cannot use only the logo.
- You must include the name and registration number of the site to which the use of the logo applies.
- You cannot use the statement of participation to advertise products.
- You cannot use the statement of participation on the products themselves.
- You cannot use the statement of participation on product packaging.

---

[19] Council Regulation (EEC) No 1836/93 of 29 June, 1993 allowing voluntary participation by companies in the industrial sector in a Community eco-management and audit scheme – *Official Journal of the European Communities* (10 July, 1993), OJ 10.7.93 L 168.

Chapter 2

The initial environmental review

Objective of the chapter

The objective of this chapter is to explain what an initial environmental review is (Figure 2.1). This chapter will provide you with the skills necessary to complete an initial environmental review, which will subsequently enable you to develop and maintain a functional EMS. After finishing this chapter, you should be able to answer the following questions:

✓ What is an initial environmental review?

✓ Why is an initial environmental review important?

✓ Why do you perform an initial environmental review?

✓ When do you perform an initial environmental review?

✓ What should be covered in an initial environmental review?

✓ What should be included in an initial environmental review report?

✓ How do you identify significant environmental aspects and impacts?

✓ What questions should you ask about the site to be reviewed?

✓ What documents should you consult to complete an initial environmental review?

✓ What must be reviewed to meet the requirements of ISO 14001 and EMAS?

✓ What written procedures are associated with the initial environmental review?

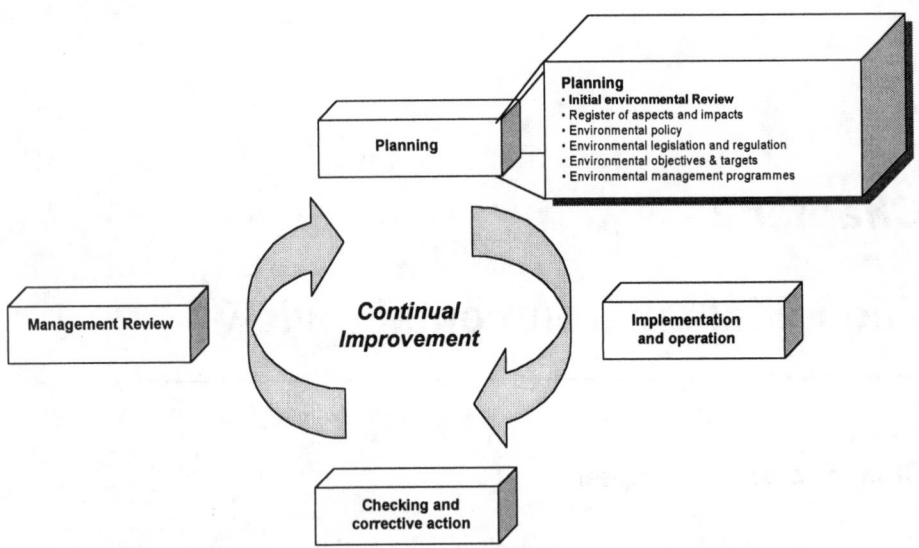

Figure 2.1 *Stages of ISO 14001 implementation: Initial environmental review*

Note

The initial environmental review is one of the more difficult tasks in developing and implementing an EMS. To help make this process easier refer to Annex 1 which contains a completed initial environmental review report case study. This case study was conducted for United Distillers' malt distilling site at Dailuaine, Scotland. It covers each of the necessary components of a complete review and should be consulted where additional information and support is needed.

Additional multimedia case studies covering the initial environmental review can be found in the handbook web site at:

http://www.entropy-international.com/handbook/

What is the initial environmental review?

Accepting that all organizations have an impact on the environment and that improved environmental performance is a direct result of controlling and minimizing those impacts, a first step towards improving performance would be to assess what impacts need to be controlled or minimized. This is the purpose of the initial environmental review (IER). It is *a systematic identification and documentation of the significant environmental impacts (or potential impacts) associated directly or indirectly with your organization's activities, products and processes.*

By completing this process of identification, you are able to improve your environmental performance by controlling the operations (activities, products and processes) that cause the significant environmental impacts identified. In short, controlling aspects reduces or minimizes impacts.

> **Note**
>
> *The initial environmental review is sometimes referred to as an initial review, a preparatory review or an environmental review but should not be confused with environmental audits or environmental impact assessments. An environmental audit or EMS audit, like a financial or quality system audit, is the process of assessing whether or not the EMS of an organization is functioning as it should by comparing it against predetermined criteria and assessing conformance to those criteria. An environmental impact assessment (EIA) is the process of assessing the potential environmental impacts (effects) of a planned activity or project, such as developing a new production process or building a dam.*

When to do an initial environmental review?

The IER is the 'preliminary snapshot' of where you are at the beginning of your EMS implementation process and should be seen as a first and fundamental step in developing, implementing and maintaining a functional EMS.

> **Note**
>
> *As mentioned in Chapter 1, an initial environmental review is an explicit requirement of EMAS and only an implied requirement of ISO 14001. However, it is strongly recommended that a formal and comprehensive IER be completed regardless of the system you choose to develop and implement.*

What should be covered in an initial environmental review?

To develop, implement and maintain a management system that effectively improves corporate environmental performance, your organization must ascertain what it is that must be improved. Consequently, an IER must identify all the significant environmental impacts of your operations. The EMS subsequently provides the framework to minimize those impacts by controlling the environmental aspects (activities, products and processes) that cause those impacts.

One very important concept here is that of *significance*. As every company, site or organization has, and always will have, some impact on the environment, it is virtually impossible to minimize simultaneously and continually all environmental impacts of your activities, products and processes. To ensure that your organization is always doing the most it can to improve continually, it is important that you are able to attach some degree of significance or priority to the environmental impacts identified in the IER.

Hint

For an EMS to be effective and certifiable to ISO 14001, it is essential that you have a clearly identified procedure to determine the significance of identified environmental impacts. As this is one of the more difficult steps in developing an EMS, the issue of significance and a procedure for determining significance are dealt with in greater detail later in this chapter (Section VI, Review of the activities, products and processes).

While there is no one set way to conduct and subsequently report the findings of an IER, there are a number of key areas that should be covered by a complete review. If these are covered sufficiently, the review will become an excellent starting point for a functional EMS worthy of ISO certification or EMAS registration.

An IER should, at the very least, cover the following four key areas.

Review of environmental management practices of the organization

If your organization is to improve its environmental performance by controlling the aspects that cause identified impacts, it will require a functional management structure to do so. Therefore, as a first step, it is crucial to establish what environmental management structure, if any, already exists. By doing this, you should be able to determine what structural management improvements would be required to effectively control the activities, products and processes that cause identified significant environmental impacts.

Review of the activities, products and processes of the organization

Once your company has established what is needed to develop a functional environmental management structure, you are ready to identify what that improved structure will manage.

Thus, the second key area to be covered in the IER is the identification of environmental aspects of your activities, products and processes (cause) that have a significant environmental impact (effect) on the environment.

Hint

The review of the activities, products and processes enables you to develop a register of environmental aspects and impacts, which is an essential component of a certifiable EMS.

Review of past environmental accidents and incidents

If your company has successfully established what environmental aspects need to be managed and what is needed to manage those aspects, the foundation of a functional environmental management system is well on its way. However, while the preceding areas will provide a clear assessment of your company's present management needs and environmental aspects and impacts, these alone will not identify the full impacts of your site. To prepare a comprehensive identification of your company's full environmental impact, it is essential to assess any present or future impacts that may result from past activities of the site on which your company operates.

Thus, the third key area to be covered in the IER is an identification of past environmental accidents and incidents that may have occurred on the site and that could lead to present or future environmental impacts.

Review of relevant legislation

One of the fundamental requirements of a certifiable EMS and continual environmental improvement is that your organization at the very least complies (or commits to comply) with all legislation, regulations, authorizations and industry codes of conduct that are associated with the environmental impacts (actual or potential) of your operations.

Thus, the fourth key area to be covered in the IER is the identification of all legislation, regulations, authorizations and industry codes of conduct that are associated with your actual or potential environmental impacts.

Hint

The review of legislation enables you to develop a register of environmental legislation and regulations, which is an essential component of a certifiable EMS.

Planning your initial environmental review

Now that you have established what information you will need to gather in your IER, it is time to plan the review itself.

> **Hint**
>
> *Remember, the quality of your review will be directly proportional to the quality of your planning!*

> **Note**
>
> *If you are a full-time employee within the company you plan to review please keep in mind that some of the points made in this section may not apply to you.*

Whether you are planning a review or a holiday, to complete the review successfully or enjoy your holiday there are always practical issues to be considered. Whether it is ensuring you have your passport or catching the right train, good planning of the physical activities of your review will have a great effect on the work you are able to do and the quality of the information you are able to gather and report.

> **Hint**
>
> *Success can be defined as the science of being totally prepared!*

Planning: The review team

To conduct a review efficiently, you must establish (well in advance) who will be included in the review team. While some reviews are performed by employees of the site being reviewed, that is not always the case or in fact optimal. Thus, when selecting a review team, ask yourself the following questions:

- What is the appropriate size and membership of your review team?

> **Hint**
>
> *The review should be performed by more than one person. As a rule of thumb there should be at least one team member per 100 employees at the site being reviewed, with a maximum team of four or five people.*

- Who will be the team leader?

> **Hint**
>
> *Every review team should have a leader. The leader should be chosen based on past review experience, past experience with the industry or sector being reviewed and their relationship with the company being reviewed. The leader should be ultimately responsible for the review, its planning, preparation and execution, and the final review report.*

- Who will be the liaison with the company before and after the review is complete (i.e. when you and your team are off-site and need answers or information)?

> **Hint**
>
> *To avoid confusion it is best that only one person from your team be selected to liaise with the company during the preparation stage.*

- How will you divide the work among your team when you get to the site?

- Does your team have sufficient practical, technical and logistical expertise to conduct the review?

- Who will be responsible for logistics such as transportation, accommodation, materials required etc.?

- Will you need translators or interpreters when working with the employees on the site?

Planning: The site to be reviewed

Just as there are a number of practical things you must tend to when selecting the review team, to complete the review itself successfully you must consider a number of issues about the site you are going to review. In planning the review, you should ask yourself the following questions:

- Where is the site and how long will it take your team to get there?

- How big or small is the site?

- Is the site spread out or in one place?

- How many people work at the site and how many of them will you need to interview to get sufficient information about the aforementioned four key areas to be covered by the review?

> **Note**
>
> *You should try to interview the site or general manager, the production/process manager and at least one person from each of the main process areas on site.*

- Given the number of interviews and complexity of the process, how much time will you need on site?

> **Hint**
>
> *The on-site part of the review should normally take between three and five days.*

A typical review schedule for a site of 200 employees with between six and eight main processes would be as follows:

Day 1: Morning	Introductions and preliminary site walk-through.
Day 1: Afternoon	Review of environmental management practices: including interviews with key personnel in this area and gathering of supporting documents.
Day 2: Morning	Review of activities, products and processes (first half of operations): including interviews with key personnel in this area and gathering of supporting documents.
Day 2: Afternoon	Review of activities, products and processes (second half of operations): including interviews with key personnel in this area and gathering of supporting documents.
Day 3: Morning	Review of past environmental accidents and incidents and review of legislation.
Day 3: Afternoon	Wrap up and double-checking to ensure all necessary information has been gathered.

- Will your team have a place to work at the site?

> **Hint**
>
> *It is important that the review team has a place to work at the site. This should be an area where the team can spread material about and discuss things without disturbing others. A separate meeting room or office is optimal.*

- Do you have access to a printer, photocopier, telephone, fax machine, e-mail etc. if needed?

Planning: Things you may need

Initial environmental reviews are sometimes conducted a long way from your own workplace and at some distance from local amenities. Thus, it is very important that you bring with you all the equipment and supplies that may be needed to complete the review. When planning what to bring some questions you may ask yourself are as follows:

- What supplies and equipment do you need?

- Do you need cameras, dictaphones, batteries, computers, stationery, printers etc.?

> ### Hint
>
> *As you will gather a tremendous amount of information during a review, facts may be forgotten, confused or misinterpreted when the time comes to prepare the off-site review report. To ensure that the information gathered is remembered and reported correctly, it is a very good idea to make use of recording equipment such as cameras, video recorders and dictaphones whenever possible.*

- Will your work be affected by the weather or the site itself (i.e. do you need rain gear, special shoes, warm clothing, light clothing)?

- Will you need any additional material during the review (dictionaries, standards, regulations, legislation, industry-specific best practice information, sector application guides etc.)?

Planning: Review logistics

While you may have a very good idea of what information you may need to present an excellent review report, if the activities required to accomplish your information-gathering process are poorly organized, the review and subsequent report will be compromised. When planning the review logistics some questions you may ask yourself are as follows:

- Where is the review to be conducted?

- When is the review to be conducted and what are the normal working hours of the personnel you will be working with?

> **Hint**
>
> *It is very important to get an understanding of the environmental aspects and impacts associated with start-up and shut-down times as well as normal working hours.*

- How will you get there?

- Who is the company liaison to whom you can address logistics questions?

- What information can you obtain in advance?

- Who at the site will be responsible for you, your team and your various requirements?

- Who will provide you with a site walk-through?

- What is your anticipated work schedule during your site visit and is that suitable to the site being reviewed and the people on site you will need to work with?

> **Hint**
>
> *It is not very useful to plan a work day of 9am-5pm if the site operates between 6am and 2pm!*

- How much time will you need on site?

> **Note**
>
> *For a review of three days, you will require at least one full day of a site employee's time for things like the site walk-through, document gathering and introductions to other site personnel.*

- Who will you need to interview and are they going to be available while you are there (see *Conducting the initial environment review* below)?

- Where will you stay if it is an overnight excursion?

Conducting the initial environmental review

Understandably, your initial environmental review report will become the synthesis of hundreds if not thousands of facts, figures, findings and recommendations. As your review report will become the first building block of your future (or improved) EMS, the information it contains must be presented in an understandable, logical and useful manner.

Consequently, the value of your review report will be dependent on a number of factors, such as how well you conduct the review, what information you gather, how you present the information gathered and the recommendations you generate. To ensure that the review process is effective, that the information gathered is useful and that the final report can actually serve as the foundation of an EMS, it is important to have a clear idea from the outset what information you will present and how that information will be presented.

Just as there is no one set way to conduct an initial environmental review, there is no one set way to report the findings of a review. However, as you can see from the IER report in Annex 1, one methodology is to gather and organize information according to the way that information is to be presented in the IER report itself.

Therefore the review should be conducted with a view to presenting the following:

I.   Table of contents

II.   Executive summary

III.   Introduction to the initial environmental review

IV.   Overview and general information

V.   Review of existing environmental management practices

VI.   Review of activities, products and processes

VII. Review of past environmental accidents and incidents

VIII. Review of relevant environmental legislation, regulations, authorizations and industry codes of practice

IX.   Recommendations for improvement

X.   Supporting information

Hint

The information does not necessarily have to be gathered in this order but the report itself should follow this format.

I. Table of contents

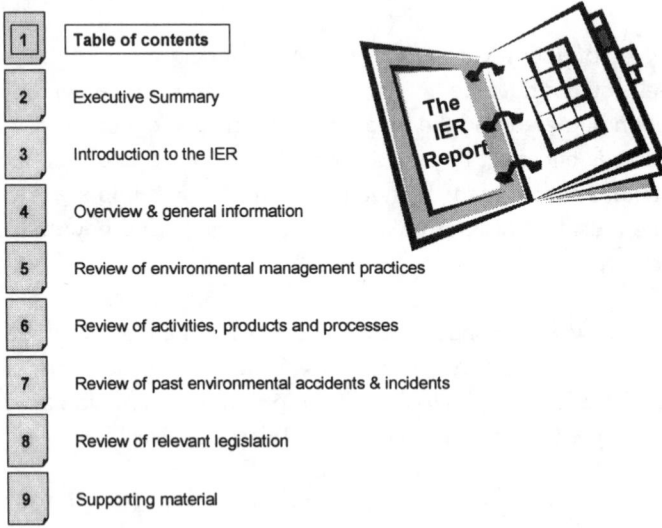

1	Table of contents
2	Executive Summary
3	Introduction to the IER
4	Overview & general information
5	Review of environmental management practices
6	Review of activities, products and processes
7	Review of past environmental accidents & incidents
8	Review of relevant legislation
9	Supporting material

Like any comprehensive report, your review report will need to be accompanied by a table of contents to clearly show readers how the report information is presented. The report should be divided into sections with headings and page numbers and the table of contents should make clear reference to each of these.

II. Executive summary

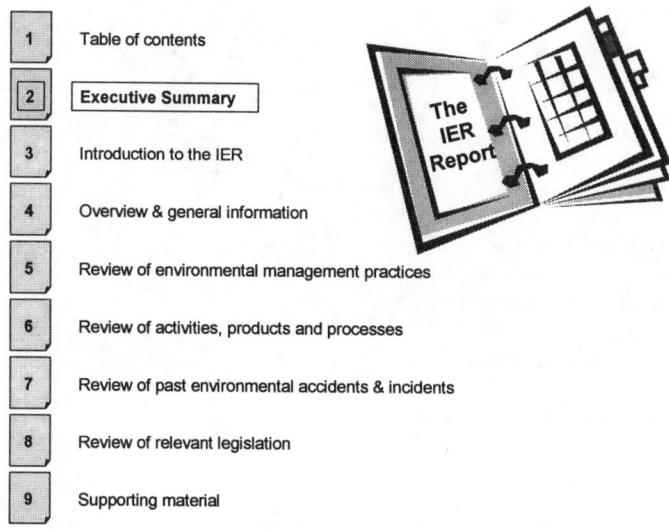

As your review will most likely contain a tremendous amount of information and as not all readers will need to know all details contained within the report, it is important to prepare an executive summary that clearly presents the main findings and recommendations of the report and review itself. These main findings and recommendations should be based on the issues of significance in the four key areas covered by the review.

Your executive summary should begin with a brief introduction and subsequently list:

- The main recommendations for improved environmental management practices

- The main recommendations for improved environmental performance of the activities, products and processes, including a list of the significant and notable aspects and impacts identified

- The main recommendations based on past environmental accidents and incidents at the site

- The main recommendations from review of environmental legislation and regulations

---

Hint

You should write your executive summary last, when the rest of the report is complete and you have determined what it is that should be summarized.

III. Introduction to the initial environmental review

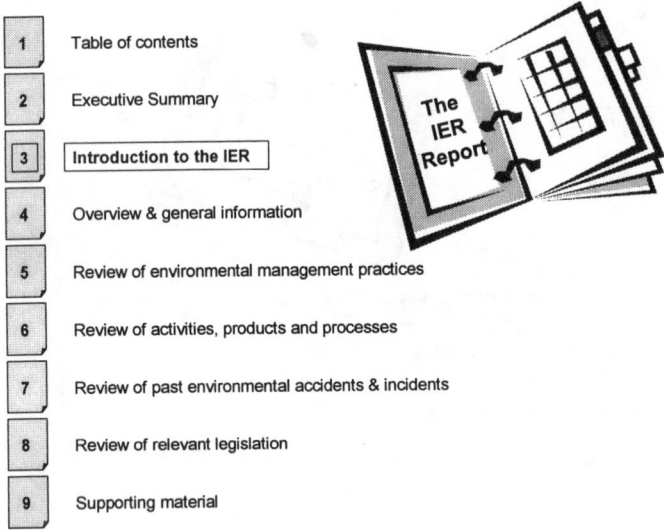

Like any introduction, the introduction to the review report should provide background information about the review and report. Essentially, your introduction should address the following:

- Purpose of the review

- Scope of the review

- Methodology used to gather the information presented

- Project team

- Time-frame for the review

Purpose of the review

In describing your purpose, you should try to answer the following questions:

- What is the purpose of the review?

- Why has the review been carried out?

> **Hint**
>
> *If possible, state exactly why the organization has asked you to conduct the review. For example, is it the starting point for an EMS as specified by ISO 14001 or EMAS or is it perhaps the starting point for a waste minimization programme or a cleaner production/pollution prevention programme?*

Scope of the review

In describing your scope, you should try to answer the following questions:

- What area, site and processes will the review cover?

- What are the physical boundaries of the review?

> **Hint**
>
> *If you were to draw a boundary around your work and work area, where would that boundary be, what would be inside it and is there anything of significance that is not?*

- What does the review not cover and what are the limitations of the review?

> **Hint**
>
> *Most reviews will not cover absolutely everything on site. For example, a cafeteria run by an external subcontractor may not be covered in your IER. If something is not covered, you should make that clear from the start to avoid confusion over the completeness of your final results.*

Methodology used to gather the information presented

To provide the reader of your review report with confidence in the information you present, it is good practice to explain how that information was gathered and where it has come from. Therefore, in describing your methodology, you should answer the following question:

- How was the information gathered (i.e. did you look at documents, interview people, complete questionnaires etc.)?

> **Note**
>
> *Use the names of people and their positions where possible to clarify the interviews that were conducted.*

Project Team

In describing your project team, you should answer the following questions:

● Who was the team leader?

● Who was in the review team?

Time-frame of the review

As the environmental impacts of a company and site change over time and as an initial review should be used as the first yardstick by which you measure future progress, it is very important to establish when your initial environmental review was conducted. In describing your time-frame, you should try to answer the following questions:

● When was the review conducted?

● How long did it take?   *3 days per comp the r~~*

> ### Note
>
> *As it is you and your review team that know most about the review report itself (to which the preceding questions apply) you will likely be able to provide most of the information in the introduction without having to look very far or interview company personnel. However, to complete the remainder of the review and gather information on the following sections, you will have to either interview site personnel or consult company documents.*
>
> *To make this process easier you will find printable check-lists for each of the following sections in Annex 2 and on the floppy disk provided in the back of this handbook, and at the web site, http://www.entropy-international.com/handbook/.*

IV. Overview and general information

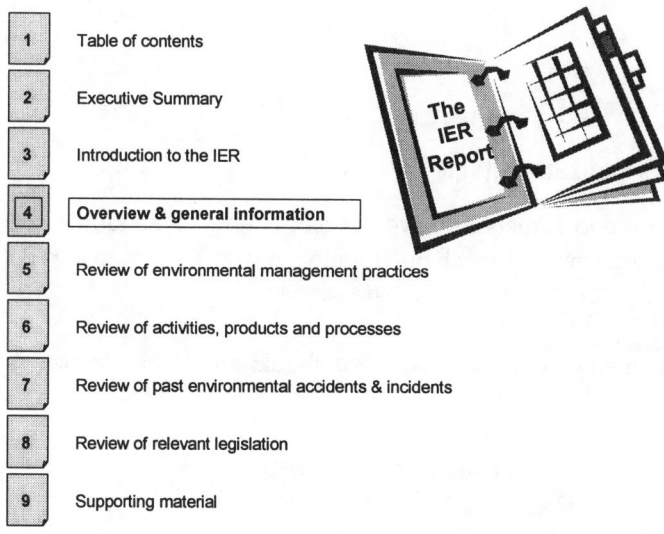

1. Table of contents
2. Executive Summary
3. Introduction to the IER
4. Overview & general information
5. Review of environmental management practices
6. Review of activities, products and processes
7. Review of past environmental accidents & incidents
8. Review of relevant legislation
9. Supporting material

Having introduced the review report itself – before plunging into detail about environmental management practices, environmental aspects and impacts, or legislation – the next step is to introduce briefly the company site or sites being reviewed.

As your review could be read by people not directly associated with the site you have reviewed (i.e. a head office management representative, the environmental manager of a separate branch or a certifier), it is important to 'introduce' the company as you would to someone who may not already know anything about it.

When describing the site being reviewed you should be sure to include the right quantity and quality of information. In simplest terms, you should try to introduce the reader to the site and highlight the issues that play, or could play, a significant role in the present and future environmental management of that organization.

Overview and general information should address the following:

- The company or organization itself

- The site

- Location of the site in relation to risk receptors and surroundings

- Topography, hydrology and geography of the site

- Other local industry

- Site history

The company or organization itself

This section of your report is meant to give the reader an idea about the company being reviewed. Is it big, small, a branch organization, a limited company or sole trader? Do they make cars, toys, energy or provide services?

In describing the company being reviewed, you should answer the following questions:

- What is the name of the company being reviewed?

- Are there parent or subsidiary companies associated with this organization?

- What is the ownership structure?

- How old is the company and how long has it been operating at its present location?

- What is the company culture (i.e. proactive, reactive, leaders, followers etc.)?

- Are they industry/trade association members; if so which associations?

- What are the main activities, products or services of the company?

The site

Having understood who is being reviewed and what it is the company does, it is important to understand where these identified activities take place.

In describing the site, you should answer the following questions:

- What is the physical location of the site being reviewed (i.e. address, town, country etc.)?

- How big is the site being reviewed (1 hectare, 5 square kilometres, 8 square miles)?

- How many employees are there at the site?

- Is the site located in one area or is it spread out?

- What is the physical appearance of the site (i.e. is it well kept, clean and organized or is it unsightly and messy)?

Hint

Site appearance may at first not seem very important, but as corporate environmental improvement is achieved by improved control and organization, a site that is disorganized and unsightly may say a lot about the present state of environmental affairs.

Location of the site in relation to risk receptors and surroundings

Having described 'who, where and what' about the company being reviewed, it is now important to highlight issues that could play a more important role in the present or future environmental management at the site. In this section, you should try to explain what environmental significance, if any, the immediate surroundings may have on environmental management at the site.

In describing the site in relation to risk receptors and surroundings, you should try to answer the following questions:

- Are there areas of natural significance in the vicinity (national parks, sensitive ecosystems, spawning/breeding grounds, wetlands, endangered species etc.)?

- Are there areas of cultural/historical significance in the vicinity (burial grounds, archaeological sites etc.)?

- Is there residential housing nearby?

- Are there any schools, hospitals, public parks, arenas or public attractions nearby?

- Is there a prevailing wind in the area?

- What is the land use to the north of the site?

- What is the land use to the east of the site?

- What is the land use to the south of the site?

- What is the land use to the west of the site?

Hint

Will the significance of an environmental impact be increased or decreased due to the location of the site?

Topography, hydrology and geography of the site

In this section, it is important to try and begin to assess where environmental aspects will have their impact. Assessing the key characteristics of the site's surroundings will help you to pinpoint areas of potential significance later on in the review process.

In describing the site's topography, hydrology and geography, you should try to answer the following questions:

- Could the site's actual or potential environmental aspects be exaggerated or mitigated by physical surroundings?

- Is the site in a valley, on a flood plain, on a hillside etc.?

- Is the site near a river, stream, sea, lake, estuary etc.?

- Have any previous hydrological or geological studies been undertaken at the site?

- Are there wells, aquifers or springs nearby?

- Is the area susceptible to seismic activity (earthquakes, tremors etc.)?

- If there are spills, leaks or uncontrolled discharges, where are they going to go?

- Is there potential for contaminating water sources?

Other local industry

An organization's environmental aspects and impacts can have an effect on neigh-bouring organizations (often catalytic) and vice versa. Therefore, it is crucial to assess what industries have the potential to both affect and be affected by your organiza-tion's environmental aspects.

In assessing how other local industries may affect or be affected by your company's aspects, you should try to answer the following questions:

- Are there other industries or companies located nearby?

- Are they located upwind or downwind of your site?

- Are they located upstream or downstream of your site?

- Are they using the same watercourses as your site (up and/or downstream)?

- Does your company co-operate in any manner with these other industries or companies?

- Has your company ever received complaints from, or complained to, these other industries or companies?

- Does your company know what these other industries or companies produce, emit and use in their operations?

- Do these other industries or companies know what is produced, emitted or used by your site?

- Do your site's environmental aspects and impacts have an obvious effect on any other neighbouring organizations?

- Do the environmental aspects or impacts of neighbouring organization's affect your site?

- Is there the possibility that your site's environmental aspects and impacts are mit-igated or exaggerated by the environmental aspects and impacts of a neighbouring site?

Site history

To develop a clear picture of present and future environmental aspects and impacts, it is necessary to establish what activities preceded yours on site. The main purpose of this section is to explain what possibility there is that the site may be the source of an environmental impact associated with past activities or products of the site. To do this you will need to develop an understanding of activities that preceded yours on site and subsequently ascertain what possibility there is that the site may have unrecognized environmental impacts associated with any previous activities identified.

In assessing the site history, you should try to answer the following questions:

- What activities, if any, have preceded present activities on the site?

- Is it possible that a previous owner or occupier has contaminated the site?

- Is there the possibility that your company's environmental aspects and impacts are mitigated or exaggerated by past activities on the site?

V. Review of environmental management practices

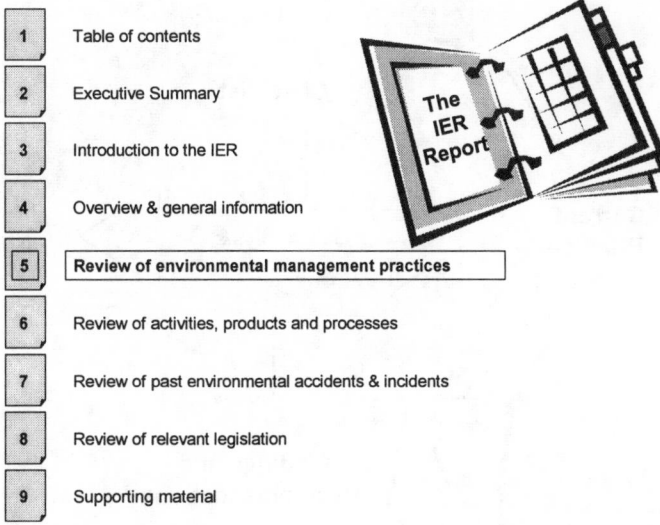

1	Table of contents
2	Executive Summary
3	Introduction to the IER
4	Overview & general information
5	Review of environmental management practices
6	Review of activities, products and processes
7	Review of past environmental accidents & incidents
8	Review of relevant legislation
9	Supporting material

As you can see from Figure 2.2 on the next page, the main objective of the review of environmental management practices is to:

- Describe the current environmental management practices within the organization.

- Identify where the current management practices do not meet the requirements of the EMS desired (gap analysis).

- Assess the magnitude of the gap between current and desired environmental management practices (findings).

- Develop recommendations for improvement based on the gaps identified.

> **Note**
>
> *The desired EMS would be that described in the purpose of the review itself: i.e. 'an EMS that conforms to the requirements of ISO 14001'.*

Your review of environmental management practices will vary depending on the size, nature and scale of your company's operations. However, the type of information you gather and the manner that information is presented should not differ in the final review report.

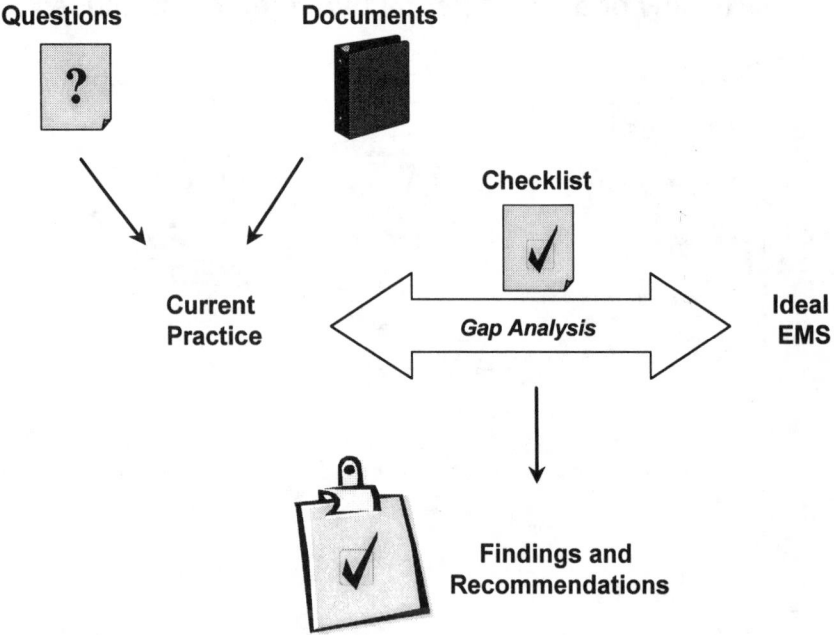

Figure 2.2 *Conducting a gap analysis*

The management areas that should be reviewed to provide a comprehensive analysis of current environmental management practices are as follows:

1 Overall management description (in general terms)

2 Environmental policy

3 The register of environmental aspects and impacts

4 Environmental legislation and regulations

5 Objectives and targets

6 Environmental management programmes

7 Environmental structure and responsibility

8 Training, awareness and competence

9 Environmental communication

10 EMS documentation

11 Environmental document control

12 Operational control

13 Emergency preparedness and response

14 Monitoring and measurement

15 Nonconformance, correction and prevention

16 Environmental records

17 EMS audits

18 Management reviews

Hint

You may notice that these 18 management areas correspond directly to the subsequent chapters of this handbook and the requirements of an EMS certifiable to ISO 14001. Essentially, after completing the initial environmental review, the remainder of the EMS implementation process (and the remainder of this handbook) is devoted to the development of each of these 'management system' components. When developed, implemented and maintained, these components (the management system itself) will facilitate the control of significant environmental aspects and impacts identified in the remainder of the review process.

Description of current environmental management practices

Having identified what a review of environmental management practices should cover, the steps outlined below will provide a useful guide for gathering, organizing and presenting information to describe current environmental management practices at your site.

To describe current environmental management practices, complete the following steps.

Step 1: Prepare questions to ask

In this step, prepare a list of questions (from the templates in Annex 2 or from the floppy disk included in the back of this handbook) whose answers will provide you with necessary information to describe current environmental management practices at the site.

Step 2: Identify documents to consult

Prepare a list of documents (from the list in Annex 2 or from the floppy disk included in the back of this handbook) that you wish to consult to provide information about current environmental management practices.

Note

The questions and documents suggested are by no means fixed and should be seen as a starting point to which you should add your own questions to be asked and documents to be consulted. Keep in mind many of the questions and documents suggested could be used to generate information for Sections VI to VIII of the IER process.

Step 3: Describe current environmental management practices

Prepare a list identifying the people you wish to interview and prepare a schedule indicating when you hope to interview the people identified. Perform the interviews and consult requested documents to generate findings of current environmental management practices.

Hint

The list of questions to be asked, documents to be consulted and a proposed interview schedule should be provided to the relevant people in advance to allow for their feedback and provide them with enough time to prepare for the questions and interviews proposed.

Preparing the gap analysis

Having described the current state of environmental management affairs, it is possible to compare this current state against that of the EMS desired at your site. This process is known as a *gap analysis* and should assess what needs to be done to meet the management system requirements. This can be done by comparing current environmental management practices against those of the ISO 14001 standard itself.

Note

The International Standard ISO 14001 can be obtained by consulting your national representative of ISO. For contact information, see Annex 4.

To develop a gap analysis against the ISO 14001 standard, complete the following step.

Step 1: Identify where the current management practices do not meet the requirements of the EMS desired (gap analysis)

Using the check-lists provided in the *Are you ready?* sections of subsequent chapters (also provided in printable form on the floppy disk in the back of the handbook or at the Internet web site), assess the gap between current environmental management practices and the environmental management practices required of an EMS conforming to ISO 14001.

Generate findings

Step 1: Describing the results of the gap analysis

Use the results of your gap analysis to develop 'findings' for each of the 18 sections discussed above. As the 18 *Are you ready?* check-lists contain hundreds of questions in total, it is likely that you will only wish to develop a finding where an improvement to current management practices is required.

Hint

Findings are essentially your description of the 'gap' between current and desired management practices.

Recommendations for improved environmental management practices

Step 1: Develop recommendations

Having described current environmental management practices and assessed the gap between current and desired environmental management practices (findings), it is now possible to develop recommendations for the areas of the management system that need improving to conform to the requirements of ISO 14001. Recommendations should be brief and to the point and should describe *actions* required to get your current environmental management practices closer to the requirements of your desired EMS (ISO 14001).

Note

Recommendations are your suggestions about what has to be done to 'narrow the gap' between current and desired environmental management practices. Your recommendations should be prioritized. The rating of priority does not need to be complicated and could be done using a simple scale, such as: 1 = critical, 2 = important, 3 = notable.

Step 2: Report preparation

Present the information generated in the preceding steps. The information should be presented in the following order:

1   Current environmental management practices (description)

2   Gap analysis (check-list results)

3   Findings (description of gap between current and desired environmental management practices)

4   Recommendations (prioritized list of actions that would narrow the gap between current and desired environmental management practices)

> **Note**
>
> *The results of your management review can be reported in a number of ways although the presentation of the gap analysis results first provides an attractive visual representation of your assessment at a glance.*

VI. Review of the activities, products and processes

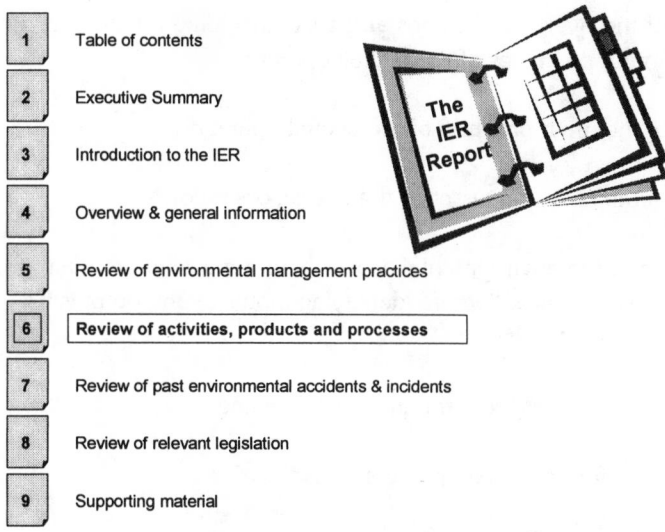

1. Table of contents
2. Executive Summary
3. Introduction to the IER
4. Overview & general information
5. Review of environmental management practices
6. **Review of activities, products and processes**
7. Review of past environmental accidents & incidents
8. Review of relevant legislation
9. Supporting material

Having reviewed the environmental management practices of your company and established what EMS components are required in your company, it is next necessary to assess what impact the activities, products and processes have on the environment. Essentially, having expressed what is needed for a functional management system, it is now possible to assess what will be managed to improve your environmental performance.

The main objective of the review of the activities, products and processes is to:

- Determine what environmental impacts are caused (or could be caused) by the site's activities, products and processes (aspects).

- Determine which of those identified impacts are significant.

The *review of activities, products and processes* and the *review of past environmental accidents and incidents* can be completed using the Significance Wizard – a simple, user-friendly drag-and-drop software program that can be obtained from Entropy International or at the web site, http://www.entropy-international.com/sigwiz/index.htm.

While your review of activities, products and processes will vary according to the size, nature and scale of your company or site, the information you gather and the manner in which it is presented in the final review report should not vary. In all cases, a complete review of activities, products and processes should include the following:

1   A description of the overall operations at the site including a flowchart diagram identifying the main processes of the overall operation

2   A description of the main products of the overall operation

3   A description of the main processes of the overall operation

4   An identification of the environmental aspects and impacts of the main processes including a mass balance diagram identifying inputs to and outputs from each of the main processes (findings)

5   A significance test for all aspects and impacts identified

6   Recommendations for each main process step

As most sites are made up of a myriad of processes, steps, departments and interactions, this part of the IER can often be a complex and daunting task. However, it is a logical, straightforward and relatively painless process if done carefully. To ensure that your investigation is both smooth and comprehensive it is best to tackle this part of the review by visualizing the overall operation as a number of independent and sequential steps of production (or services) rather than a single production unit (Figure 2.3).

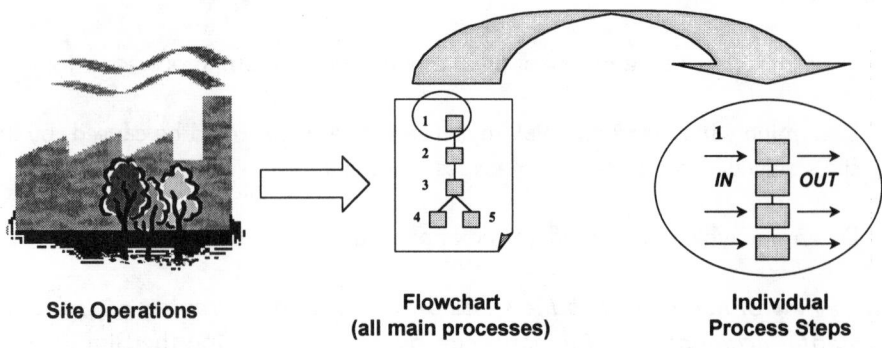

| Site Operations | Flowchart (all main processes) | Individual Process Steps |

Figure 2.3 *Creating the flowchart*

As is the case with the whole review process, there is no one set way to complete the review of the site's activities, products and processes. However, a helpful methodology for completing this part of the review is presented in the actions listed below.

Just as the IER process began with an introduction to the organization being reviewed, the review of activities, products and process should begin with an introduction to the overall site operations being reviewed. The first step, therefore, is to provide a clear picture of the overall site operations.

Description of overall site operations

The description of overall site operations should explain what is generally happening at the site. This section should be brief and to the point. Finer details about individual process steps should be addressed in later sections.

To describe the overall site operations, complete the following steps.

Step 1: Prepare questions to ask

Prepare a list of questions (from the following list or from the floppy disk included in the back of this handbook) whose answers will provide you with information to describe the overall operations at the site.

Questions to consider:

- What is the name of the overall activity or process on the site?

- What is (are) the final product(s) of the overall site operation?

- What quantity(s) is (are) produced and at what market value?

- How many main processes are there in the overall site operation and what are these processes called?

- What are the main inputs to and outputs from the overall site operation?

- Who has overall responsibility for the site?

- What are the operating hours for the site?

- Who are the major suppliers of inputs to the site?

- Who are the major customers or clients for the products of the site?

Step 2: Identify documents to consult

Prepare a list of documents (from the following list or from the floppy disk included in the back of this handbook) that you may wish to consult to give you additional information about the overall site operations.

Documents to consult:

📖 Site plan

📖 Overall site operations flowchart

📖 Area map

📖 Products list

Step 3: Generate information on overall site operations

Prepare a list identifying the people you wish to interview and prepare a schedule indicating when you hope to interview the people identified. Perform the interviews and consult requested documents to generate information on overall site operations.

> **Hint**
>
> *The list of questions to be asked, documents to be consulted and a proposed interview schedule should be provided to the relevant people in advance to allow for their feedback and provide them with enough time to prepare for the questions and interviews proposed.*

Step 4: Develop a flowchart diagram for overall site operations

Develop a flowchart that visually represents the overall operations at the site. The flowchart should clearly indicate each of the main processes of the site operations. Each 'box' of the flowchart should show each of main processes to be reviewed in the section below 'description of the main processes of overall site operations'.

> **Hint**
>
> *To make your job easier and the final result less confusing, use existing process flowchart diagrams, divisions and process names where possible. Flowcharts often exist in places like quality management manuals and training materials.*

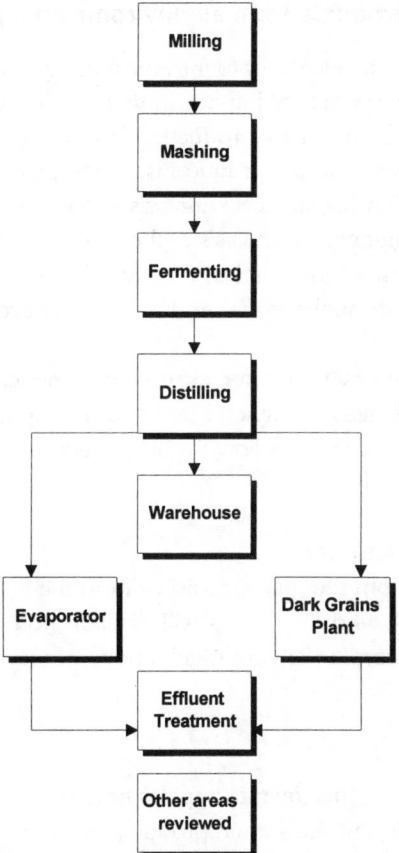

Figure 2.4 *United Distillers, Dailuaine overall process flowchart*

Hint

You will notice that the United Distillers main process flowchart (Figure 2.4) has a step entitled 'Other areas reviewed'. This is added (in almost all cases) as there are some aspects and impacts that are not associated with a single main process but many steps of the overall site operations. You will also notice that each of the main process flowcharts (discussed later) have a similar final box entitled 'Associated with all steps'. This is for the same reason and should be added to most, if not all, flowcharts developed.

Step 5: Describe the overall operations at the site

Using the information gathered from interviews, documents and general observations, prepare a description of the overall site operations.

Description of the main products from an environmental point of view

Having 'introduced' the site it is necessary to introduce the products as they too may be a significant source of environmental impact. In many cases, the greatest environmental impact of a site is not from its production but its products. Similarly, the greatest environmental impact of many products is during their use or final disposal rather than their production. A full life cycle analysis is not required for certification; however, to develop a comprehensive analysis of the total impact of an organization, it is important to assess the environmental aspects and impacts of its products both within the factory gates and throughout the product's full life cycle.

To create a full picture of the environmental aspects and impacts of an organization it is necessary to develop, at least in general terms, a description of the product(s) from an environmental point of view. Follow the steps below.

Step 1: Prepare questions to ask

Prepare a list of questions (from the following list or from the floppy disk included in the back of this handbook) whose answers will provide you with information to describe the site's products from an environmental point of view.

Questions to consider:

- Are there any significant environmental impacts associated with the extraction of the raw materials used in the site's main product(s)?

> **Hint**
>
> *Products such as furniture made from tropical hardwood or jewellery made from ivory have few environmental impacts associated with their production, use or even final disposal but have considerable impacts associated with the extraction of their raw materials from which they are made: i.e. tropical deforestation and the threatening of endangered species!*

- Are there any significant environmental impacts associated with the use of the site's main product(s)?

> **Hint**
>
> *Products such as toilets or light bulbs have significant environmental impacts associated with their raw material extraction, production or disposal but also have considerable impacts associated with their use: i.e. considerable water and energy use!*

- Are there any significant environmental impacts associated with the disposal of the site's main product(s)?

> **Hint**
>
> *Products such as refrigerators (containing CFCs and PVC plastics) have significant environmental impacts associated with their raw material extraction, production and use but also have considerable impacts associated with their final disposal: i.e. ozone depletion, air pollution and land contamination!*

- Are there any significant social impacts associated with the raw material extraction, production, use or final disposal of the site's main product(s)?

> **Hint**
>
> *Products such as cigarettes, alcohol, cars etc. have considerable social impacts that should also be considered even if not clearly identifiable in environmental terms. Remember that people are part of the environment and an impact on people is an impact on the environment.*

- Are environmental criteria considered when purchasing raw materials?

> **Hint**
>
> *You should consider materials used directly and indirectly in the products. For example, wood is used directly in the production of furniture while cleaning agents for cleaning woodworking tools are used indirectly in the production of furniture.*

- Has a life cycle analysis (LCA) been performed for any of the site's main product(s)?

- Are there any 'design for the environment' (DFE) criteria or parameters applied in the production of the site's main product(s)?

- Do any of the site's main product(s) have 'eco-labels' or belong to any 'eco-labelling schemes'?

- Is 'extended producer responsibility' considered in the design and development of the site's main product(s)?

Step 2: Identify documents to consult

Prepare a list of documents from the following list (or from the floppy disk included in the back of this handbook) that you wish to consult to provide information about the site's main product(s).

Documents to consult:

📖 Raw material purchase orders

📖 Product life cycle analysis reports

📖 Product design policies or design for the environmental parameter reports

Step 3: Describe the site's main product(s) from an environmental point of view

Prepare a list identifying the people you wish to interview and prepare a schedule indicating when you hope to interview the people identified. Perform the interviews and consult requested documents to describe the site's main product(s) from an environmental point of view.

Hint

The list of questions to be asked, documents to be consulted and a proposed interview schedule should be provided to the relevant people in advance to allow for their feedback and provide them with enough time to prepare for the questions and interviews proposed.

Step 4 (optional): Life cycle analysis

Prepare a simplified life cycle analysis of the site's main product(s). To do this you should explain each of the main phases of the product's life cycle 'from cradle to grave'. This description should, at a minimum, analyse raw material extraction, product design and development, production, product use and final product disposal.

Hint

It may be easier to do this by developing a flowchart of the product's life cycle, identifying the significant environmental aspects and impacts of each phase, step by step.

Description of the main processes of the overall site operations

Now that you have described the overall site operations and the site's main product(s), it is possible to describe the main processes within the overall site operations. This next section should briefly describe each of the main processes identified in the overall operations flowchart developed above.

Hint

As the overall site operations flowchart, and the individual process steps being explained, may only include the process steps that are directly related to the production of the main product(s) at the site, it is important also to consider the indirect process steps in your evaluation. Indirect steps would include things like maintenance, administration, construction, deconstruction, storage, catering and cafeterias. If these, or other indirect steps, are missing, you should add them to your description and subsequent analysis as though they were a main process step and complete a description, a description from an environmental point of view and a review of environmental aspects and impacts for each of them.

Like the description of overall site operations, the description of main processes should be clear and to the point.

Hint

To avoid confusion, when describing the main processes you should use the same names as used in your overall site operations flowchart.

As depicted in Figure 2.4, the main processes described in the United Distillers case study were:

- Milling
- Mashing
- Fermenting
- Distilling
- Warehouse

- Evaporation
- Dark Grains Plant
- Effluent Treatment
- Other areas reviewed

To describe each of the main process steps, follow the steps below.

Step 1: Prepare questions to ask

Prepare a list of questions (from the following list or from the floppy disk included in the back of this handbook) whose answers will provide you with information to describe each of the main processes on site.

Questions to consider:

- What are the main activities and products associated with each of the main processes?

- What is generally happening in each of the main processes?

- What is the name of each of the main processes?

- What are the individual steps within each main of the processes?

- What is (are) the product(s) of each of the main processes?

- Where on site are each of the main process steps located?

- Who is responsible for each of the main processes?

- How many employees are involved in each of the main processes?

- What are the work schedules of each of the main processes?

Step 2: Identify documents to consult

Prepare a list of documents (from the following list or from the floppy disk included) that you may wish to consult to give you additional information about the main processes on site.

Documents to consult:

- Process flowcharts

- Operations training manuals

- QMS process flowcharts

- Mass balance flowcharts

- Site maps

Step 3: Describe the main processes of the site

Prepare a list identifying the people you wish to interview and prepare a schedule indicating when you hope to interview the people identified. Perform the interviews and consult requested documents to describe the main processes of the overall site operations.

> **Hint**
>
> *The list of questions to be asked, documents to be consulted and a proposed interview schedule should be provided to the relevant people in advance to allow for their feedback and provide them with enough time to prepare for the questions and interviews proposed.*

Identification of environmental aspects and impacts associated with the main processes on site

Having identified what is generally happening at the site and in each of the main processes, it is now important to develop a clear picture of what environmental aspects and impacts are associated with each of the main processes identified.

Using the flowchart developed in the previous description of overall site operations (Figure 2.4), you should be able to separate each of the main processes identified into smaller individual process steps as depicted in Figure 2.5.

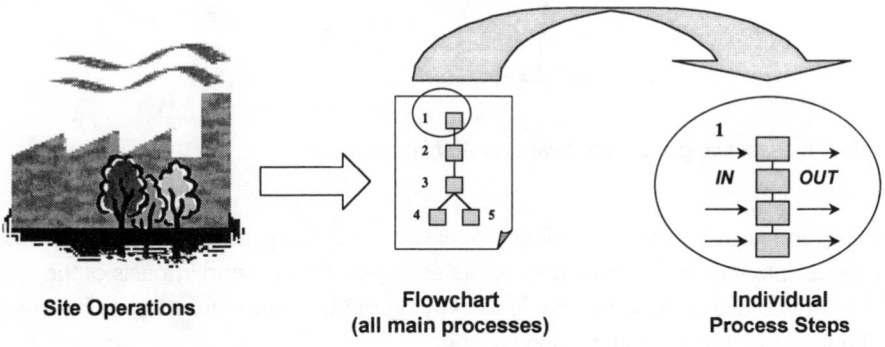

| Site Operations | Flowchart (all main processes) | Individual Process Steps |

Figure 2.5 *Creating the flowchart*

In doing so, and as shown in Figure 2.6, you will be able to assess the impacts of smaller and more manageable parts of the operation piece by piece and comprehensively identify the aspects and impacts of the site as a whole.

As you will notice from the questions in Annex 2, a comprehensive assessment of each process will include questions about:

- Water use
- Energy use
- Chemical use
- Raw material use
- Storage
- Effluents to water

- Emissions to air
- Disposal to land
- Hazardous, special or restricted substances or waste
- Abnormal situations

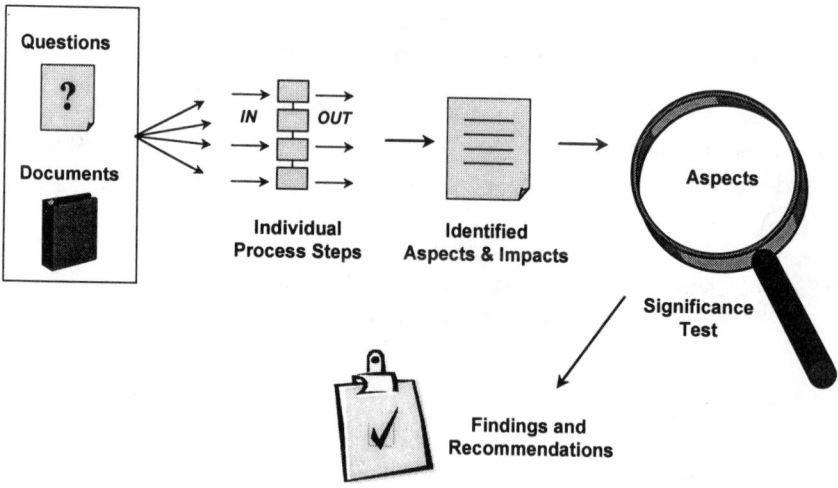

Figure 2.6 *Assessing significant aspects and impacts*

Having broken down the overall site operation into main processes, and the main processes into individual process steps, assessing the aspects and impacts of the overall site operations is done by simply identifying the inputs to and outputs from each of the individual process steps one by one.

To identify the environmental aspects and impacts associated with the main processes, follow the steps below.

Step 1: Develop a flowchart for individual process steps

For the each of the main processes identified in the overall site operations flowchart and described in the process section, draw a flowchart to represent the individual process steps within that main process. An example of this step can be seen in the United Distillers case study diagram (Figure 2.7). In this example, the first main process (Milling) is divided into six individual process steps (Malt delivery, Storage, De-stoning, Milling, Grist Storage, Associated with all steps).

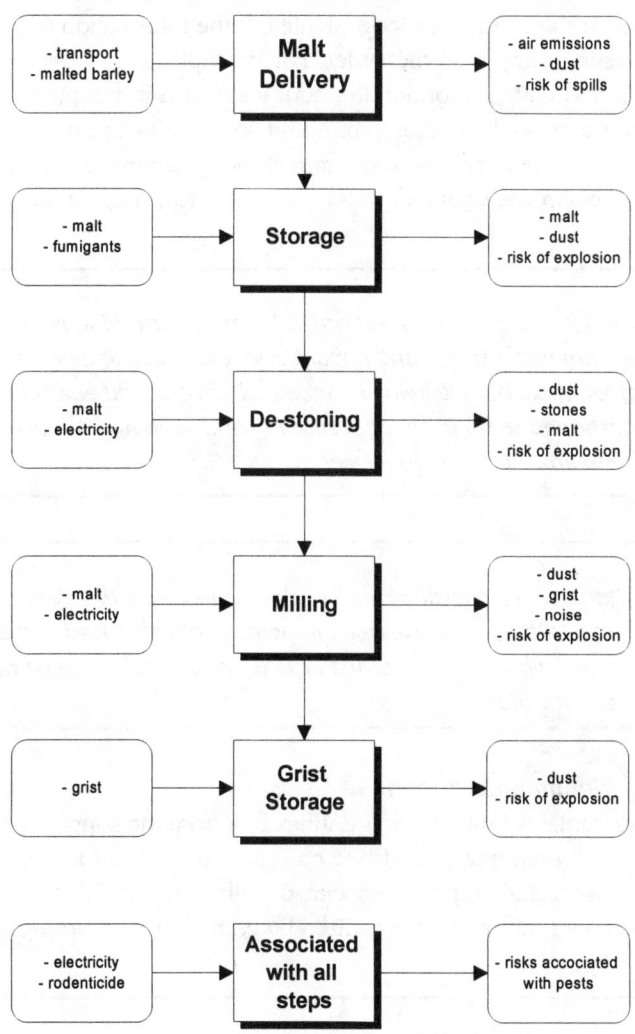

Figure 2.7 *United Distillers milling process flowchart*

Step 2: Prepare questions to ask

Prepare a list of questions (from the list in Annex 2 or from the floppy disk included in the back of this handbook) whose answers will provide you with information to describe the environmental aspects and impacts associated with each of the main processes identified in the overall operations flowchart.

To keep this information logical and organized, gather your information process by process and aspect by aspect. Deal with each main process one by one and for each main process follow the question headings below. By doing this you will provide consistent information for each main process identified; the information will similarly be gathered and presented in the same order. For example, start with the first main process identified and gather information about water use in the process, then raw material use in the process, then energy use, and so on. When you have completed all the questions for the first main process, repeat the questions for the second main process identified, beginning again at 'water use', then 'raw material use' and so on.

Hint

Be selective when choosing your questions. The important thing is to determine significant environmental aspects and impacts, not every single detail of the operations. Remember, most people being interviewed will get tired and bored quickly when being questioned so try to limit questioning and keep a watchful eye for signals of restlessness from your interviewee!

Note

If you find that one set of questions is not relevant to a process step, skip to the next set of aspect questions. For example, if there is no water used in the process, do not go on to ask what quantity of water is used, what the source of water is and what the cost of water is!

Step 3: Identify documents to consult

Prepare a list of documents (from the list in Annex 2 or from the floppy disk included in the back of this handbook) that you wish to consult to provide information about the environmental aspects and impacts associated with each of the main processes. Again, try to keep your information organized by gathering it according to aspect.

Hint

Other sections of the IER also include the gathering of documents. Check with your review team members to see where overlap occurs. Alternatively, you could prepare a long check-list of all the documents required for the entire review and account for documentation as 'needed' or 'received'.

Step 4: List the environmental aspects and impacts associated with each of the main processes (findings)

Prepare a list identifying the people you wish to interview and prepare a schedule indicating when you hope to interview the people identified. Based on the information you have gathered from interviews, documents and general observations, present your findings on the environmental aspects and impacts associated with each of the main processes. The findings should be presented in the same order in which the questions were organized and asked: i.e. water use, raw material use, energy use etc. As you can see in the United Distillers case study (Figure 2.8), the first main process (Milling) had 16 findings, the first of which was: 'Water use in this process is negligible'.

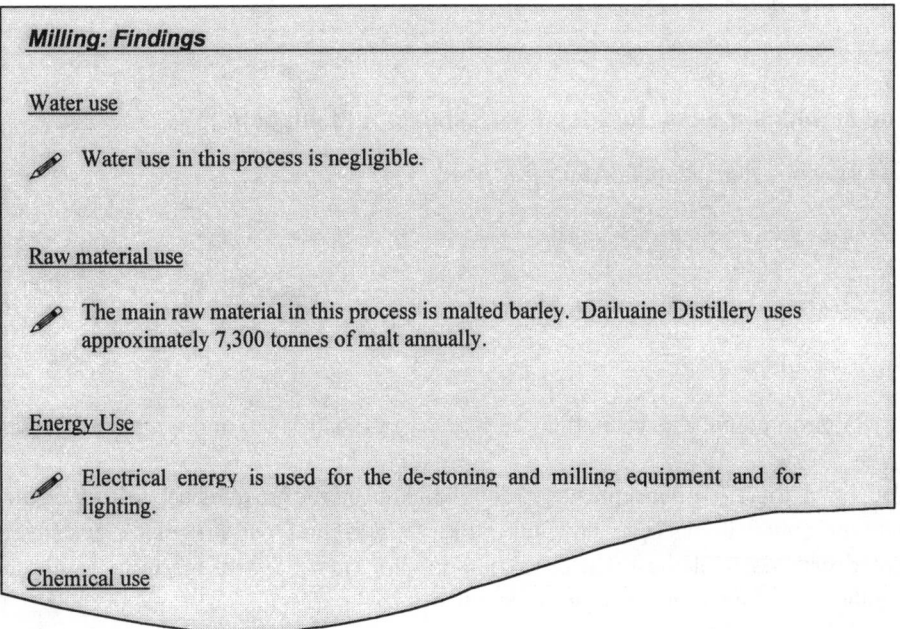

Figure 2.8 *List of findings*

Hint

The list of questions to be asked, documents to be consulted and a proposed interview schedule should be provided to the relevant people in advance to allow for their feedback and provide them with enough time to prepare for the questions and interviews proposed.

Step 5 (optional): Develop a mass balance
for the main processes identified

Develop a mass balance that visually represents, and quantifies where possible, the physical inputs and outputs for each individual process step of each of the main processes.

```
─────  Hint  ─────

The mass balance can be general but should indicate the main inputs to and out-
puts from each process step. In theory, for every process (or flowchart) there
should be a 'balance' between inputs and outputs. A mass balance is a flowchart
that visually shows inputs and outputs and enables you to identify any imbalance
between input and output and the environmental aspect or impact associated
with the inputs or outputs identified.
```

The significance test for identified aspects and impacts

The previous steps identified and described:

1   The overall activities of the site

2   The main product(s) of the site

3   Each of the main processes

4   Environmental aspects and impacts associated with each of the main processes

You have therefore systematically identified the environmental aspects associated with the entire site, its operations and products and the environmental aspects associated with each of the main processes on site. Now you are ready to assess the significance of each of the aspects identified.

This is a very important step as the aspects and correlating impacts deemed 'significant' will be the ones that are subsequently managed by you and your emerging EMS!

The reason significance is important is because *significant* aspects and impacts:

● *Must* be addressed in your environmental policy (even if only in general terms).

● *Must* be included in your register of environmental aspects and impacts.

● *Must* have objectives and targets set for them.

- *Must* have environmental management programmes established to meet those objectives and targets.

- *Must* be addressed by your training programmes.

- It is significant aspects that you *must* control to minimize your company's environmental impact and ultimately improve corporate environmental performance!

The assessment of significance will always be, to some degree, subjective and based on personal opinion, local factors, current events and so on. Therefore, to ensure that aspects and impact are evaluated with consistency, it is essential to have a documented methodology (procedure) for assessing significance. This 'significance test' is important not only because of the implications that significant aspects and impacts have on your EMS and environmental performance but it is a requirement of ISO 14001 to have a written procedure used to determine which aspects and impacts identified are deemed significant.

As all companies that have been certified to ISO 14001 have had to use some form of a significance test for identified aspects and impacts, and as this test will always be somewhat subjective, there is no one correct 'significance test'. Some are complicated and detailed while others are generic and simple. What methodology you use depends on your situation and the degree of detail with which you wish to test your identified aspects and impacts.

To help you through this task you will find below two 'significance test' methodologies that can be used. Significance Test Methodology 1 is simple and generic, while Significance Test Methodology 2 is more comprehensive and more detailed. If you intend to certify your EMS, it is recommended that you use Significance Test Methodology 2 or the Significance Wizard (which can be obtained from Entropy International or at the web site, http://www.entropy-international.com/sigwiz/index.htm).

Significance Test Methodology 1

Step 1: Identify the aspects and impacts to be tested

Before you start, create a list of the aspects and impacts that you have identified and for which you would like to assess significance. (This should have been completed in the previous section.) To keep your information organized and presentable, it is recommended that you divide the identified aspects and impacts according to the main processes with which they are associated.

Step 2: The significance test

For each identified aspect and impact ask the following questions:

1   Is the identified aspect or impact associated with any legislation, regulation, authorizations or industry codes of practice by which your site or company is bound?

2   Is the identified aspect or impact the source of complaints from employees, neighbours, stakeholders or the community in which you operate?

3   Is the identified aspect or impact of concern to your employees, shareholders, bankers, customers, clients, insurers or lawyers?

4   Is the identified aspect or impact clearly associated with any of the more serious global environmental issues such as:

   • Global warming and the greenhouse effect

   • Ozone depletion

   • Acid rain and acidification

   • Eutrophication

   • Deforestation

   • Loss of biodiversity

   • Non-renewable resource use

5   Is the identified aspect or impact associated with the use of substances that are known, or suspected to be, toxic to plant, animal or human life on the planet?

Step 3 (optional): Prioritization

To help you develop prioritized objectives, targets and management programmes, you may want to attach some degree of priority or weighting to the aspects and impacts identified as significant. One simple way of doing this is to give a value of 1 to a Yes answer and a value of 0 to a No answer. The weighting would simply be the total summation of the value attached to each aspect or impact tested.

Step 4: The results

If you answered Yes to any of these questions, the aspect or impact in question should be considered 'significant'. You should prepare a list of all the aspects and impacts identified as being significant.

Note

This methodology is based on assessing significance by using the precautionary principle. The principle simply implies that 'when in doubt, err on the side of caution and assume that the impact is significant'.

Significance Test Methodology 2

As you will notice, there is a set of blank templates and a *Significance Test User's Guide* at the end of this chapter. Use these templates, the *Significance Test User's Guide* and the *General impact description list* to complete the following steps.

Note

The Significance Wizard software program (available at the web site, http://www.entropy-international.com/sigwiz/index.htm) follows this test methodology.

Step 1: Complete Form 1 of significance test
for each of the main processes

Using the templates in the *Significance Test User's Guide* (also provided in printable form on the accompanying floppy disk at the back of this handbook), complete *Form 1: Process activity descriptions and environmental aspects* for each of the main processes described above and identified in the overall flowchart. For additional guidance, refer to the completed significance tests in the United Distillers initial environmental review report case study in Annex 1.

Step 2: Complete Form 2 of significance test
for each of the main processes

Using the templates at the end of this chapter (also provided in printable form on the accompanying floppy disk at the back of this handbook), complete *Form 2: Process activities and environmental aspects matrix* for each of the main processes of the overall flowchart (and for which you completed a Form 1). For additional guidance, refer to the completed significance tests in the United Distillers initial environmental review report case study in Annex 1.

Step 3: Complete Form 3 of significance test
for each of the main processes

Using the templates at the end of this chapter, (also provided in printable form on the accompanying floppy disk at the back of this handbook), complete *Form 3: Process environmental impact description and significance matrix* for each of the main processes of the overall flowchart (and for which you completed a Form 2). For additional guidance, refer to the completed significance tests in the United Distillers initial environmental review report case study in Annex 1.

Step 4: Prepare a list of significant environmental aspects
and impacts for each of the main processes

Having completed a significance test for each of the identified aspects of each of the main processes on site, it is now possible to generate a prioritized list of the significant environmental aspects and correlating impacts for each of the main processes.

As you will notice from the *Significance Test User's Guide*, the significance test will generate a result between 0 and 25 for each of the identified aspects or impacts tested. Any aspect or impact with a value greater than or equal to 8 is 'notable' and any aspect or impact with a value greater than or equal to 12 is 'significant'.

Your prioritized list should identify, in order of significance, the aspects and impacts that were deemed 'significant' in your significance test. You should list the significant aspects and impacts for each of the main processes and include the full list of aspects and impacts deemed significant in the executive summary. It is good practice also to include aspects and impacts identified as being notable for each of the main processes in the executive summary.

Step 5 (optional): Generate a graph showing the identified significant aspects and impacts

For each of the main processes, develop a graph showing each of the aspects and impacts identified and which of those were deemed 'significant' or 'notable' during the significance test. This step will provide you with a visual representation of the aspects and impacts that must be controlled and minimized. This format of illustrating significant aspects and impacts is very effective in presentations, board meetings and training seminars as it quickly and easily identifies those aspects of your operations that:

- *Must* be addressed in your environmental policy.

- *Must* be included in your register of environmental aspects and impacts.

- *Must* have objectives and targets set for them.

- *Must* have environmental management programmes established to meet those objectives and targets.

- *Must* be addressed in training programmes.

- *Must* be controlled to minimize your company's environmental impact and ultimately improve your corporate environmental performance!

Recommendations for improved environmental performance of the activities, products and processes

Step 1: Develop recommendations

Now that you have identified and prioritized the significant environmental aspects and impacts of the site's products and processes, it is possible to develop recommendations for improvement. Although recommendations will vary depending on the site and the products and processes being reviewed, recommendations should be:

- Based on all the information gathered in the previous sections of the review

- A synthesis of facts, findings, ideas and common-sense solutions

- Specific to the identified significant environmental aspects or impacts for both the main product(s) and the main processes described earlier in the review

Additionally, recommendations for each of the significant aspects and impacts identified should take into account information gathered to describe:

- The site

- Location of the site in relation to risk receptors

- Topography, hydrology and geography

- Other local industry

- Site history

For example, if the risk of oil spillage from a process is deemed significant and that process area is located on the bank of a river, the recommendation for that significant aspect should correspondingly reflect both the aspect and its proximity to the river.

Recommendations should be generated for:

1   The findings of the review of main product(s) from an environmental point of view

2   Each of the main processes described (and identified in the overall process flow-chart)

Note

The recommendations for each of these areas should be presented in their respective section.

Hint

Recommendations should be based on information from a variety of sources. Some recommendations may be based purely on common sense while others will require deeper investigation into a given aspect's toxicity, effect on the local ecology, or health and safety implications.

VII. Review of past environmental accidents and incidents

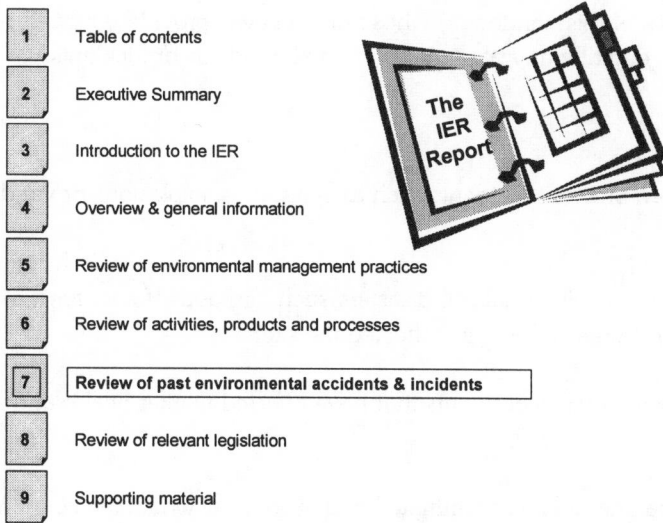

1	Table of contents
2	Executive Summary
3	Introduction to the IER
4	Overview & general information
5	Review of environmental management practices
6	Review of activities, products and processes
7	Review of past environmental accidents & incidents
8	Review of relevant legislation
9	Supporting material

By completing the preceding exercises, you will have generated ample information about the current environmental management practices and the present environmental aspects and impacts of your organization. However, for a complete IER and an EMS that will enable an organization to improve its corporate environmental performance, you will also need to address one special area from which significant environmental aspects and impact may arise – the past.

Note

If you recall, Section IV (Overview and general information) addressed the issue of site history. This earlier section was intended to create a picture of the environmental accidents and incidents that may have occurred before your occupation of the site. This next section, 'Review of past environmental accidents and incidents', is intended to create a picture of the environmental accidents and incidents that have occurred during your occupation of the site.

To develop an understanding of the significant environmental aspects and impacts that may be associated with your site's past, complete the following steps.

Step 1: Prepare questions to ask to describe any past environmental accidents or incidents

In this step, prepare a list of questions (from the following list or from the floppy disk included in the back of this handbook) whose answers will provide you with necessary information to describe any past environmental accidents or incidents.

Question to consider:

- Have there been previous incidents such as spills, fires, explosions or vandalism on the site?

- Have there been previous natural disasters such as floods, forest fires, storms, earthquakes or droughts that have affected the site?

- Have there been previous accidents that have affected worker health and safety on site?

- Is it possible that the site is contaminated and, if so, what would the contaminant most likely be?

- Have there been previous audits performed; if so, did the results of the audit indicate any past accidents or incidents?

- Have there been internal complaints about your company's past environmental performance? If so, what were the complaints about?

- Have there been external complaints about your company's past environmental performance? If so, what were the complaints about?

- Have local authorities performed any site inspections; if so, were their findings of past accidents or incidents?

- Have independent companies or consultants performed any site inspections; if so, were their findings of past accidents or incidents?

- Has your company been fined or warned for past regulatory/legislative noncompliance? If so, what was the nature of the noncompliance?

- Have there been previous wastewater system backups, overflows or failures?

- Have there been incidents of accidental or uncontrolled discharge from the site in the past?

- Have there been any accidents, spills, leaks etc. involving stored materials on site in the past; if so, what material?

- Have there been any accidents, spills, leaks etc. with restricted and hazardous materials on site in the past; if so, what material?

Step 2: Identify documents to consult to describe any past environmental accidents or incidents

Prepare a list of documents from the following list (or from the floppy disk included in the back of this handbook) that you wish to consult to provide information about past environmental accidents or incidents.

Documents to consult:

📖 Previous complaints forms

📖 Past notices of legislative nonconformance

📖 Record of fines, injunctions, court cases etc. involving the site

Step 3: Describe past environmental accidents or incidents (findings)

Prepare a list identifying the people you wish to interview and prepare a schedule indicating when you hope to interview the people identified. Perform the interviews and consult requested documents to generate findings on past environmental accidents or incidents.

> ### Hint
>
> *The list of questions to be asked, documents to be consulted and a proposed interview schedule should be provided to the relevant people in advance to allow for their feedback and provide them with enough time to prepare for the questions and interviews proposed.*

Step 4 (optional): Complete a significant test for identified past environmental aspects and impacts

While it is important to assess the relative significance of all identified aspects and impacts, those identified in this section will have already occurred; the recommendations will generally not be affected by a significance test.

However, if you wish to complete a significance test for the aspects and impacts identified, simply follow the steps outlined in one of the significance tests discussed above.

Recommendations based on past environmental accidents and incidents at the site

Step 1: Develop recommendations
Having assessed what environmental accidents and incidents may have occurred on site since current operations began, it is important to generate recommendations based on your finding.

As discussed above in Section VI (*Review of the activities, products and processes*), recommendations should be a synthesis of facts, findings, ideas and common-sense solutions and should be specific to the environmental aspects or impact identified.

Similarly, recommendations for each of the aspects and impacts identified in this section should take into account information gathered to describe:

- The site

- Location of the site in relation to risk receptors

- Topography, hydrology and geography

- Other local industry

- Site history

VIII. Review of relevant environmental legislation, regulations, authorizations and industry codes of practice

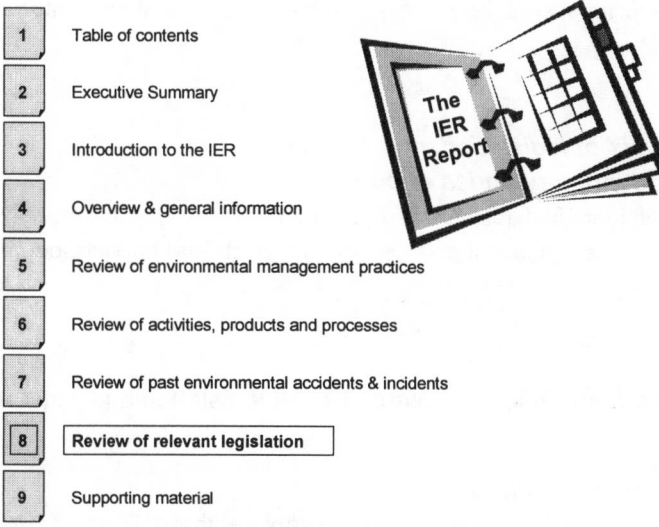

1. Table of contents
2. Executive Summary
3. Introduction to the IER
4. Overview & general information
5. Review of environmental management practices
6. Review of activities, products and processes
7. Review of past environmental accidents & incidents
8. **Review of relevant legislation**
9. Supporting material

Having now completed a review of past environmental accidents and incidents, there is one last but key area that should be addressed within the initial environmental review – the review of environmental legislation, regulations, authorizations and industry codes of practice. As this section is also covered in Chapter 5, the description here will be kept brief.

This review involves the evaluating of all legislation, regulations, authorizations and industry codes of practice associated with either your environmental aspects or your environmental impacts.

To review relevant environmental legislation, regulations, authorizations and industry codes of conduct, complete the following steps.

Step 1: Identify all the legislation, regulations, authorizations and industry codes of practice that apply to your operations

As it is a requirement of ISO 14001 to identify and have access to legislation and regulations relevant to your significant environmental aspects, unless your company has a well-established legal department it is best to consult a commercial legislation database or register right from the start. Such databases and registers have been prepared in many countries, often save considerable amounts of time and are updated regularly. For contacts to such service providers see the appropriate links in the handbook web site at http://www.entropy-international.com/handbook/.

Other sources of information on relevant legislation and regulations can be found through industry associations, all levels of government, your local authority or your local regulator.

Step 2: Gap analysis of legislation, regulations, authorizations and industry codes of conduct

Compare the list of identified relevant legislation, regulations, authorizations and codes of practice with the significant environmental aspects and impacts identified in the previous sections.

Recommendations from review of environmental legislation and regulations

Step 1: Develop recommendations

Having assessed what relevant legislation, regulations, authorizations and industry codes of conduct are relevant to your operations, and assessed which of those you may not be compliant with, it is possible to generate recommendations for improvement. Quite simply, if it has been determined that you are in a situation of noncompliance, a recommendation to rectify the situation is required.

As simple as this process sounds, it should not be taken lightly and all areas of non-compliance should be considered significant and dealt with accordingly.

IX. Supporting material

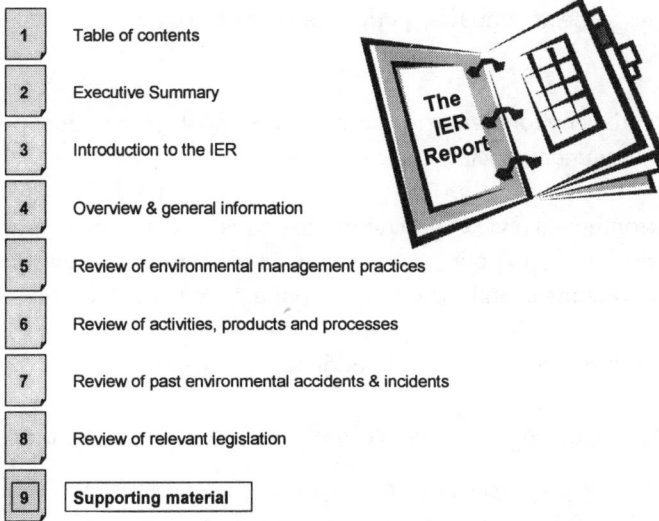

1. Table of contents
2. Executive Summary
3. Introduction to the IER
4. Overview & general information
5. Review of environmental management practices
6. Review of activities, products and processes
7. Review of past environmental accidents & incidents
8. Review of relevant legislation
9. **Supporting material**

Although you have now gathered all the information that would usually constitute a standard IER, there are often circumstances when supporting information would add value to your report. This sort of information is best added as annexed material and should be identified in the table of contents.

Such information could include:

- A glossary of terms if unfamiliar or complex terms are being used

- Photographs of significant aspects and impacts

- Literature to support recommendations

- Lists of important contacts

- Electronic versions of the report

- Surveys and statistics

- Questionnaires completed or used during the review

- Site and process maps

- Permit details

- Monitoring records, etc.

Recommendations for successful EMS implementation

✓ Your initial environmental review should be a comprehensive analysis of environmental issues, aspects, impacts, performance and controllable activities of your organization.

✓ Performing an initial environmental review should establish your current position in relation to environmental performance.

✓ Your initial environmental review should form the basis for developing your organization's environmental policy, objectives and targets, environmental management programmes, and the other components of the entire EMS.

✓ Your initial environmental review should address:

- Your company's current environmental management practices and procedures

- Environmental aspects (causes) and impacts (effects) associated with your organization's activities, products and processes

- Past environmental accidents and incidents

- All the relevant environmental legislation, regulations, authorizations and industry codes of conduct to which your organization subscribes

✓ More specifically, your initial environmental review should identify:

- All inputs to your activities, products or processes

- All outputs from your activities, products or processes

- All air emissions (controlled and uncontrolled) from your activities, products or processes

- All effluents (controlled and uncontrolled) from your activities, products or processes

- The generation or disposal of solid and other waste (particularly hazardous waste) associated with your activities, products or processes

- Any contamination of land as a result of your activities, products or processes

- All uses of raw materials and natural resources associated with your activities, products and processes

- All other discharges or emissions associated with your activities, products or processes such as thermal energy, noise, odour, dust, vibrations and visual impact

- All environmental issues of local or community relevance associated with your activities, products or process and any issues associated with your company and its environmental performance

✓ Your initial environmental review should identify impacts arising from:

- Normal activities, products and processes

- All or any abnormal activities, products and processes

- All accidents and potential emergency situations associated with your activities, products and processes

- All past, present and planned activities, products and processes

- The life cycle of your products

Significance test templates and User's Guide

Three blank templates are given below, followed by the *Significance Test User's Guide*.

Follow the *User's Guide* to help complete the templates for significant aspects and impacts.

Form 1

Process activity descriptions and environmental aspects

Site:

Main process:

Date:

Individual process steps	Reference	Description of individual process steps	Aspects	
			Normal conditions	Abnormal conditions

Form 2
Process activities & environmental aspects matrix (page 1)

Site: **Date:**

Process:

Aspect reference numbers	General aspects	Process steps 1 2 3 4 5 6
WU	**Water Use**	
WU01	Use of water from municipal sources	
WU02	Use of water from surrounding water courses	
WU03	Other water use	
EU	**Energy Use**	
EU01	Use of natural gas (not including use for transportation)	
EU02	Use of oil (not including use for transportation)	
EU03	Use of coal (not including use for transportation)	
EU04	Use of other fossil fuels (not including use for transportation)	
EU05	Use of fuel for transportation	
EU06	Use of energy from nuclear-generated sources	
EU07	Use of energy from hydro-generated sources	
EU08	Use of energy from wind-generated sources	
EU09	Use of energy from solar-generated sources	
EU10	Use of electricity from mixed sources	
EU11	Other energy use	
CU	**Chemicals Use**	
CU01	Use of restricted chemicals	
CU02	Use of acidic chemicals (not on list of restricted chemicals)	
CU03	Use of basic chemicals (not on list of restricted chemicals)	
CU04	Use of solvents (not on list of restricted chemicals)	
CU05	Use of hydraulic oils, lubricants etc.	
CU06	Other chemical use	
RU	**Raw Materials Use**	
RU01	Use of raw materials (hazardous, special or restricted)	
RU02	Use of raw materials (not hazardous, special or restricted)	

Form 2
Process activities & environmental aspects matrix (page 2)

Site: **Date:**

Process:

Aspect reference numbers	General aspects	Process steps 1 2 3 4 5 6
RU	**Raw Materials Use (continued)**	
RU03	Use of packaging material (not including RU01 or RU02)	
RU04	Use of office materials (not including RU01, 02, or 03)	
RU05	Use of construction materials (not including RU01, 02, 03 or 04)	
RU06	Other raw material use (not including RU01, 02, 03, 04, or 05)	
ST	**Storage on Site**	
ST01	Storage of chemicals	
ST02	Storage of raw materials	
ST03	Storage of hazardous, restricted or special substances	
ST04	Storage of waste (not hazardous, restricted or special)	
ST05	Storage of hazardous, restricted or special waste	
ST06	Other storage	
EW	**Effluents to Water**	
EW01	Discharge of effluent to treatment facility	
EW02	Controlled discharge of treated effluent to water course	
EW03	Controlled discharge of untreated effluent to water courses	
EW04	Uncontrolled discharge of treated effluent to water courses	
EW05	Uncontrolled discharge of untreated effluent to water courses	
EW06	Discharge of hazardous, restricted or special effluent	
EW07	Other discharges	
EA	**Emissions to Air**	
EA01	Emission of process gases/heat within the process (not up flue)	
EA02	Emission of flue gases/heat (not including NO_x, SO_x, particulates)	
EA03	Emission of NO_x	
EA04	Emission of SO_x	

Form 2
Process activities & environmental aspects matrix (page 3)

Site: **Date:**

Process:

Aspect reference numbers	General aspects	Process steps 1 2 3 4 5 6
EA	**Emissions to Air (continued)**	
EA05	Emission of CO_2	
EA06	Emission of particulate matter (fly ash)	
EA07	Emission of dust or raw materials from within the process	
EA08	Emission of VOCs	
EA09	Emission of hazardous, restricted, special substances (not VOCs)	
EA10	Emissions from transport	
EA11	Other emission	
DL	**Disposal to Land**	
DL01	Disposal to municipal landfill	
DL02	Disposal to site landfill	
DL03	Disposal to incineration	
DL04	Disposal to recycling, reclamation or re-use	
DL05	Disposal of hazardous, restricted or special substances	
DL06	Previous soil contamination (actual and potential for)	
DL07	Other disposal	
OT	Other	
OT01	Vibrations	
OT02	Noise, smell	
OT03	Visual impact (include lights)	
OT04	Other	
AB	Risk of Abnormal Activity	
AB01	Risk of fire or explosion	
AB02	Risk of spillage, leakage or uncontrolled discharge	
AB03	Risk of spill etc. of hazardous, restricted or special substances	
AB04	Risk to worker health and safety	
AB05	Other abnormalities	

Form 3
Process environmental impact description and significance matrix

Site:

Main process:

Date:

Process steps	Aspect or impact identified	Ref. No.	Impact Description	Direct or indirect	Impact rating	Severity rating	Significance factor

Significance Test User's Guide

Completing Form 1:
Process activity description and environmental aspects

Step 1: Fill in you company name, site and date in the **Site:** and **Date:** positions. Do so for all three forms. For additional help, refer to the United Distillers Form 1 in Annex 1. For example: *Dailuaine Malt Distillery, Scotland* and *May 1997.*

Step 2: Fill in the name of the main process (as identified in the overall flowchart) being reviewed, in the **Main Process:** position. Do so for all three forms. For example: *Milling.*

Step 3: Fill in the names of the individual process steps of the main process in the **Individual process steps** column and do so for each of the individual process steps identified as part of the main processes named. For example: the Milling process was comprised of *Malt Delivery; Storage; De-stoning; Milling; Grist Storage; All Process Steps.*

> **Note**
>
> *You will notice that the last step in Form 1 was All Process Steps. It is advisable to add this as a final step in each of the process steps to cover aspects that are associated with the main process being described but not unique to one individual process step. This would include things like lighting for the entire process, cleaning if it is for the entire process, storage etc.*

Step 4: If you have an existing management system (such as a quality management system or a health and safety system) and you intend to integrate the two management systems, begin that process here by including any corresponding process step reference numbers/names for each process step in the **Reference** column.

Step 5: In the **Description of individual process step** column, describe the process step named in the **Individual process steps** column. Do this for each process step named. For example: *The process starts with the delivery to the site of malted barley. The barley arrives by lorry in 24-ton loads and is poured into a ground level holding area where it is stored for later use in the process.*

Step 6: In the Aspects – Normal Conditions column, list the environmental aspects associated with the individual process step described in the corresponding Description of individual process step column. For example: *Use of raw material (malted barley); transportation; dust emissions.*

Step 7: In the Aspects – Abnormal Conditions column, indicate any possible abnormal, accidental or dangerous environmental aspects associated with the process step described in the Description of individual process step column. For example: *Risk of malt spillage; health risk associated with pigeons attracted by spills.*

Step 8: Complete a Form 1 for all main processes and then move on to Form 2: *Process activities & environmental aspects matrix.* You should complete a Form 1 for each of the main processes identified in the overall operations flowchart and for which you provided a description in Section VI (*Review of the activities, products and processes*).

Completing Form 2:
Process activities and environmental aspects matrix

Step 1: In the Process steps columns, fill in the names of each of the identified process steps indicated in the Individual process steps column in Form 1. For additional help, refer to the United Distillers Form 2 in Annex 1. For example: *Malt Delivery; Storage; De-stoning; Milling; Grist Storage; All Process Steps.*

Step 2: For each of the process steps identified in the Process steps columns, indicate all the aspects associated with that individual process step by marking an X in the box that corresponds to that aspect. This should be done by sequentially moving down the list of General Aspects and marking an X in the process step column if that aspect has been identified on Form 1 for that process step.

> **Note**
>
> *The number of aspects marked in the column for each process step should match exactly the number of aspects identified in the Aspects – Normal Conditions and Aspects – Abnormal Conditions columns. For example in the United Distillers case study, the first process step (Malt Delivery) should have six Xs for the six identified aspects of that process step (four Normal and two Abnormal).*

Step 3: Complete a Form 2 for all main processes and move on to Form 3: *Process environmental impact descriptions & significance matrix.* You should complete a Form 2 for each of the main processes identified in the overall operations flowchart and for which you completed a Form 1.

Completing Form 3:
Process environmental impacts description and significance matrix

Step 1: In the Process steps column, write the name of the first individual process step identified in the Individual process steps column from Form 1. For additional help, refer to the United Distillers Form 3 in Annex 1. For example: *Malt Delivery.*

Step 2: In the Aspect identified column, describe the first aspect identified in that process step. This is the aspect from the Aspects column on Form 1. For example: *Use of fuel for transportation.*

Step 3: In the Aspect ref. column, write in the Aspect Reference Number, as identified in Form 2, that corresponds to the first aspect identified in that process step. Do this for each of the aspects identified and complete this process for each of the individual process steps of that main process. For example: in the Malt Delivery, the Aspect Reference Numbers would be EU05, RU02, EA07, EA11, AB02 and AB06. These will have been marked with an X on Form 2. The quantity of Xs should exactly equal the number of ASP references for each process step.

Step 4: In the Impact Description column, and using the List of General Environmental Impact Descriptions in Annex 3, describe the environmental impact associated with that aspect. For example: The impact description for EU05 would be: *Depletion of non-renewable fossil fuel resources. Global oil resources estimated to be 70-80 years. Combustion leads to emission of VOCs, NOx, SOx, CO_2 and thus air pollution, greenhouse gas production and global warming.*

Step 5: In the Impact column, indicate the impact 'value' for each aspect identified. To do this, ask the following questions for each aspect identified. A Yes answer should be valued as a 1 and a No answer should be valued as a 0. Calculate the value (between 0 and 5) for the Impact rating column in Form 3.

Questions to ask:

1 Is the aspect associated with any legislation, regulations, authorizations or industry codes of practice? Or does the identified aspect involve the use of any hazardous, restricted or special substances?

2 Is the aspect of concern to stakeholders?

   i.e. • Employees   • Neighbours   • Bankers
         • Clients   • Shareholders   • Insurers
         • Customers   • Lawyers   • Local community

3 Is the identified aspect or impact clearly associated with any of the more serious global environmental issues?

   i.e. • Global warming and the   • Eutrophication
          greenhouse effect   • Deforestation
         • Ozone depletion   • Loss of biodiversity
         • Acid rain and acidification   • Non-renewable resource use

4 If the aspect identified is quantifiable, is the *amount* of use significant?

5 If the aspect identified is quantifiable, is the *amount* or *frequency* of use significant?

Step 6: In the Severity column, indicate the perceived severity value for each aspect identified. Using the list below, decide what rating the identified aspect should be classified as. This should reflect the effect that the aspect has or could have if uncontrolled.

Severity rating matrix	
Rating	**Severity**
1	No or minor environmental effect
2	Slight environmental effect
3	Moderate environmental effect
4	Serious environmental effect
5	Disastrous environmental effect

> ┌─ *Hint* ──
> *You may want to use half values when assessing severity. For example, you may assign a 3.5 for severity when the impact is more than a moderate concern but slightly less than a serious environmental impact. This also allows you to attach more weight to certain issues that should be addressed.*

Step 7: In the Significance Factor column, indicate the significance of the aspect by multiplying the impact and severity. For example, in the United Distillers case study: EU05 has a significance of 8.75 in this process step.

> ┌─ *Hint* ──
> *It will be easier to complete the impact, severity and significance evaluation for each aspect one by one rather than assess impact for all aspects, then severity for all aspects and so on.*

Step 8: *(optional)* To illustrate your findings graph the results to create a Process Environmental Aspects & Impacts Graph.

Are you ready?

	Yes	Partly	No
# EMS Check-list ### Initial Environmental Review *page 1 of 2*	■	◩	□
Was your initial environmental review a comprehensive analysis of the environmental issues, aspects, impacts, performance and controllable activities of your organization?	❏	❏	❏
Has your initial environmental review established your company's current position in relation to environmental performance?	❏	❏	❏
Have the findings of your initial environmental review formed the basis for developing your environmental policy, objectives and targets, environmental management programmes and the other components of the entire environmental management system?	❏	❏	❏
Has your initial environmental review assessed:			
• Existing environmental management practices and procedures in your organization?	❏	❏	❏
• All the environmental aspects and impacts associated with your organization's activities, products and processes?	❏	❏	❏
• All past environmental accidents, emergencies and incidents and regulatory noncompliance?	❏	❏	❏
• All the relevant environmental legislation, regulation and other requirements to which your organization subscribes?	❏	❏	❏
Has your initial environmental review identified:			
• All inputs to your organization's activities, products or processes?	❏	❏	❏
• All outputs from your organization's activities, products or processes?	❏	❏	❏
• All air emissions (controlled and uncontrolled) from your activities, products or processes?	❏	❏	❏
• All effluents (controlled and uncontrolled) from your activities, products or processes?	❏	❏	❏

EMS Check-list
Initial Environmental Review

page 2 of 2

	Yes	Partly	No

- The generation or disposal of solid and other waste (particularly hazardous waste) associated with your activities, products or processes? Any contamination of land as a result of your organization's activities, products or processes? ☐ ☐ ☐

- All uses of raw materials and natural resources associated with your activities, products and processes? ☐ ☐ ☐

- All other discharges or emissions associated with your activities, products or processes, such as thermal energy, noise, odour, dust, vibrations and visual impact? ☐ ☐ ☐

- All environmental issues of local or community relevance associated with your activities, products or process and any issues associated with your company and its environmental performance? ☐ ☐ ☐

Has your initial environmental review identified impacts arising from:

- Normal activities, products and processes? ☐ ☐ ☐

- All or any abnormal activities, products and processes? ☐ ☐ ☐

- All accidents and potential emergency situations associated with your activities, products and processes? ☐ ☐ ☐

- All past, present and planned activities, products and processes? ☐ ☐ ☐

- The life cycle of your organization's products? ☐ ☐ ☐

Chapter 3

The register of environmental aspects and impacts

Objective of the chapter

The objective of this chapter is to explain what the register of environmental aspects and impacts is. This chapter will provide you with the skills necessary to develop your own register of environmental aspects and impacts and ensure that the register is sufficient to maintain a functional EMS. After finishing this chapter, you should be able to answer the following questions:

✓ What is the register of environmental aspects and impacts?

✓ Why is the register of environmental aspects and impacts important?

✓ What must the register of environmental aspects and impacts include to meet the requirements of ISO 14001 and EMAS?

✓ What written procedures are required for the register of environmental aspects and impacts?

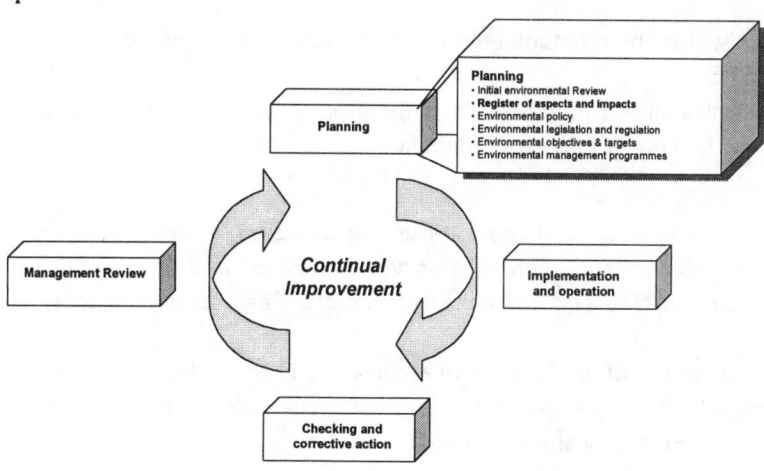

Introduction to the register of environmental aspects and impacts

As you will recall from Chapter 2, the initial environmental review provided a framework for the evaluation of your operations to identify what significant environmental impacts are associated with the activities, products and processes.

As discussed in Chapters 1 and 2, the key to improving environmental performance continually is to control the aspects of your company's operations (activities, products and processes) that have an impact on the environment. Logically, then, improved environmental performance is a direct result of managing the environmental aspects and impacts identified in the initial environmental review.

However, before an organization can 'manage' (and subsequently control and minimize) its environmental aspects and impacts it must first identify and document what those aspects and impacts are and record the findings of such a process. In short, *the register of environmental aspects and impacts is the documented record of significant environmental aspects and correlating impacts that the organization must control and minimize to improve its overall corporate environmental performance.*

This register will likely resemble parts of the executive summary of your initial environmental review. The register of environmental aspects and impacts should list all significant environmental aspects and impacts and indicate where they occur in the overall process. Essentially, the register is an account of the significant aspects and impacts:

- Addressed in the environmental policy (even if only in general terms)

- For which objectives and targets should be set

- For which environmental management programmes should be developed

- That should be addressed in training programmes for those personnel directly associated with the aspect identified

You must have a written procedure for assessing the environmental aspects and impacts associated with your activities, products and processes (see Chapter 2, Section IV, *The significance test for identified aspects and impacts*).

Your register of environmental aspects and impacts should be documented. This register should be considered a controlled document and be kept in your environmental management manual.

As you will notice in the United Distillers example, their register of aspects and impacts lists, among other things, their significant environmental aspects in order of priority. In this case, their register starts with:

Significant aspect identified	Process	Significance factor
Risk to worker health and safety: asphyxiation	Mashing & Fermentation	17.5
Risk of water contamination	All steps	16
Use of radioactive sources	Dark Grains Plant	16
Use of natural gas for fuel	Dark Grains Plant	16
Risk of explosion	Distillation	16
Disposal of copper to soil	Effluent Treatment	16
Risk of uncontrolled effluent discharge	Effluent Treatment	16
Use of powdered hydrated lime	Effluent Treatment	16
Use of radioactive sources	Evaporation	16
Discharge of cooling water to burn	Evaporation	16
Risk of explosion	Warehouse	16
Risk of explosion	Dark Grains Plant	15
Use of cooling water from burn	Distillation	15
Dust emissions	Milling	15
Risk of explosion	Milling	15
Risk of explosion	Milling	15
Risk of oil spills, leaks, discharge	Storage	15

Hint

Creating a register is a very efficient way of keeping track of your environmental aspects and impacts. For significant environmental aspects and impacts you may want to consider referencing your environmental management programmes directly from your register (for example, an identified significant aspect for energy use can be linked directly to an energy programme.)

> ┌─── **Certification tip** ───
>
> *When applying for certification, your certifier will be looking for documented evidence that you have evaluated your environmental aspects and impacts and have determined their significance. Time spent getting this stage right can help to ease the assessment (validation) process and will lead to a smoother evaluation. It is critical that you establish clear links between your identified significant environmental impacts, their corresponding aspects and the rest of your system (policy, objectives and targets, management programmes).*

As with most of the EMS requirements, the definitions of and requirements for a register of aspects and impacts differ slightly between ISO 14001 and EMAS.

In ISO 14001, the register of environmental aspects and impacts is not clearly defined (although environmental aspects and environmental impacts are) and it is not an explicit requirement for certification. It is implied, however, that an organization must set its environmental objectives for improvement based on identified environmental aspects and their correlating impacts. To this end, it is suggested in ISO 14004 (*General guidelines on principles, systems and supporting techniques*) that the results of an initial environmental review should be documented[1].

In EMAS, the register of aspects and impacts is termed 'environmental effects registration' and is described as the registration of 'the environmental effects of a company's activities at the site'. All EMAS-registered companies 'must compile a register of those environmental effects identified as significant'[2].

More precise requirements for a register of environmental aspects and impacts are discussed in the next section, *Recommendations for successful EMS implementation*.

To develop your own register you can use the blank and sample templates later in this chapter or on the floppy disk at the back of this book. For examples of what other companies have done, visit the handbook web site at http://www.entropy-international.com/handbook/.

When you think you are finished, you can audit your performance and assess your success by using the check-list in the *Are you ready?* section at the end of this chapter.

---

[1]   *ISO 14004 Environmental Management Systems – General guidelines on principles, systems and supporting techniques* (Geneva: International Organization for Standardization, 1996).

[2]   Council Regulation (EEC) No 1836/93 of 29 June, 1993 allowing voluntary participation by companies in the industrial sector in a Community eco-management and audit scheme. *Official Journal of the European Communities*, 10 July, 1993 (OJ 10.7.93 L 168).

Recommendations for successful EMS implementation

✓ Subsequent to the completion of your initial environmental review, you *must* compile a register that includes all aspects and impacts identified as being significant.

✓ This register *must* identify (where significant):

- All inputs to your activities, products or processes

- All outputs from your activities, products or processes

- All air emissions (controlled and uncontrolled) from your activities, products or processes

- All effluents (controlled and uncontrolled) from your activities, products or processes

- The generation or disposal of solid and other waste (particularly hazardous waste) associated with your activities, products or processes

- Any contamination of land as a result of your activities, products or processes

- All uses of raw materials and natural resources associated with your activities, products or processes

- All other discharges or emissions associated with your activities, products or processes such as thermal energy, noise, odour, dust, vibrations and visual impact

- All environmental issues of local or community relevance associated with your activities, products or process and any issues associated with your company and its environmental performance

✓ Your register *must* identify the significant aspects and impacts arising from:

- Normal activities, products and processes

- All or any abnormal activities, products and processes

- All accidents and potential emergency situations associated with your activities, products and processes

- All past, present and planned activities, products and processes

- The life cycle of your products

✓   Your register *must* be reviewed regularly and revised accordingly.

✓   Your register *must* be documented and presented in a clear, concise and easy-to-understand format.

✓   Your register *must* differentiate between direct aspects and impacts – those which your organization has a high degree of control over – and indirect – those which your organization does not have a high degree of control over.

✓   Your register *must* describe the procedure employed to identify environmental aspects and impacts and their significance (see Chapter 2, Section IV, *The significance test for identified aspects and impacts*).

✓   Your register *should* be kept in your environment management manual.

Register of Environmental Aspects and Impacts				*Blank template*
Company Name: **Department/Site:** **Updated by:** **Approved by:**		**Document Version:** **Issue/Revision Date:** **Replaces Version:** **Page** **of**		
Environmental aspect	Aspect reference numbers	Direct or indirect	Main activity, product or process affected	Significance factor

Register of
Environmental Aspects and Impacts

Sample template

Company Name: United Distillers   **Document Version:** RAI001V1
Department/Site: Dailuaine Distillery   **Issue/Revision Date:** 30/05/97
Updated by: Wirral Green   **Replaces Version:** None
Approved by: Grant Fromage   **Page** 1 **of** 2

Environmental Aspect	Aspect reference numbers	Direct or indirect	Main activity, product or process affected	Significance factor
Risk to WH & S: asphyxiation	AB05	Direct	Mashing & Fermentation	17.5
Risk of water contamination	AB05	Direct	All steps	16
Use of radioactive sources	RU01	Direct	Dark Grains Plant	16
Use of natural gas for fuel	EU01	Direct	Dark Grains Plant	16
Risk of explosion	AB01	Direct	Distillation	16
Disposal of copper to soil	DL05	Direct	Effluent Treatment	16
Risk of uncontrolled effluent discharge	AB02	Direct	Effluent Treatment	16
Use of powdered hydrated lime	RU02	Direct	Effluent Treatment	16
Use of radioactive sources	RU01	Direct	Evaporation	16
Discharge of cooling water to burn	EW03	Direct	Evaporation	16
Risk of explosion	AB01	Direct	Warehouse	16
Risk of explosion	AB01	Direct	Dark Grains Plant	15
Use of cooling water from burn	WU02	Direct	Distillation	15
Dust emissions	EA07	Direct	Milling	15
Risk of explosion	AB01	Direct	Milling	15
Risk of explosion	AB01	Direct	Milling	15
Risk of oil spills, leaks, discharge	AB02	Direct	Storage	15
Effluent of cooling water to burn	EW02	Direct	Distillation	14
Effluent of cooling water to burn	EW02	Direct	Distillation	14
Still vent vapour	EA01	Direct	Distillation	14
Discharge of cooling water to burn	EW03	Direct	Evaporation	14
Use of process water from burn	WU02	Direct	Mashing & Fermentation	14
Effluent of cooling water to burn	EW03	Direct	Mashing & Fermentation	14
Use of cooling water from burn	WU02	Direct	Mashing & Fermentation	14
Emission of CO_2	EA05	Direct	Mashing & Fermentation	14
Emission of VOCs	EA08	Direct	Mashing & Fermentation	14
Use of solvent-based paints	CU04	Direct	Cask Preparation	14
Emission of NOx	EA03	Direct	Dark Grains Plant	14
Emission of CO_2	EA05	Direct	Dark Grains Plant	14
Emission of VOCs	EA08	Direct	Dark Grains Plant	14

Register of
Environmental Aspects and Impacts

Sample template

Company Name: United Distillers	**Document Version:** RAI001V1
Department/Site: Dailuaine Distillery	**Issue/Revision Date:** 30/05/97
Updated by: Wirral Green	**Replaces Version:** None
Approved by: Grant Fromage	**Page** 2 **of** 2

Environmental Aspect	Aspect reference numbers	Direct or indirect	Main activity, product or process affected	Significance factor
Use of cooling water from burn	WU02	Direct	Distillation	14
Emission of still vapours	EA08	Direct	Distillation	14
Use of cooling water from burn	WU02	Direct	Evaporation	14
Risk of spills, leaks etc.	AB02	Direct	Evaporation	14
Emission of VOCs	EA08	Direct	Warehouse	14
Emission from transport	EA11	Direct	Draff Delivery	12.5
Noise	OT02	Direct	Draff Pressing & Mixing	12.5
Noise	OT02	Direct	Dryer	12.5
Noise	OT02	Direct	Milling	12.5
Dust emission	EA07	Direct	Milling	12.5
Noise	OT02	Direct	Pelletizer	12.5
Emission from transport	EA11	Direct	Sludge Disposal	12.5
Emissions from transport	EA11	Direct	Transportation of Casks	12.5
Use of mixing water from burn	WU02	Direct	Warehouse	12.25
Use of water for cleaning	WU02	Direct	All Distillation Steps	12
Use of water for cleaning	WU02	Direct	All Evaporation Steps	12
Risk of spills, leaks	AB02	Direct	All steps: Mash. & Ferm.	12
Discharge spent lees to treatm'nt plant	EW01	Direct	Distillation	12
Use of water for cleaning	WU02	Direct	Mashing & Fermentation	12
Emissions from transportation	EA01	Direct	All steps	12
Use of natural gas	EU01	Direct	Boiler house	12
Emission of flue gas	EA02	Direct	Boiler house	12
Noise	OT02	Direct	Boiler house	12
Emission from transport	EA11	Direct	Delivery of Effluent	12
Noise	OT02	Direct	Distillation	12
Noise	OT02	Direct	Evaporation	12
Transport emissions	EA11	Direct	Milling	12
Dust emissions	EA07	Direct	Milling	12
Use of rodenticide	CU06	Direct	Milling	12
Use of fumigants	CU01	Direct	Storage	12

Are you ready?

<table>
<tr><td rowspan="3">

EMS Check-list
Register of Environmental Aspects & Impacts
page 1 of 2
</td><td>◼</td><td>◩</td><td>◨</td></tr>
<tr><td>Yes</td><td>Partly</td><td>No</td></tr>
</table>

	Yes	Partly	No
Has your organization compiled a register of aspects & impacts identified as significant?	☐	☐	☐

Does your organization's register of aspects and impacts include (where significant):

	Yes	Partly	No
• All inputs to your organization's activities, products or processes?	☐	☐	☐
• All outputs from your organization's activities, products or processes?	☐	☐	☐
• All air emissions (controlled and uncontrolled) from your activities, products or processes?	☐	☐	☐
• All effluents (controlled & uncontrolled) from your activities, products or processes?	☐	☐	☐
• The generation or disposal of solid and other waste (particularly hazardous waste) associated with your activities, products or processes?	☐	☐	☐
• Any contamination of land as a result of your activities, products or processes?	☐	☐	☐
• All uses of raw materials and natural resources associated with your activities, products or processes?	☐	☐	☐
• All other discharges or emissions associated with your activities, products or processes (thermal energy, noise, odour, dust, vibrations and visual impact)?	☐	☐	☐
• All environmental issues of local or community relevance associated with your activities, products or processes and any issues associated with your company and its environmental performance?	☐	☐	☐

EMS Check-list
Register of Environmental Aspects & Impacts
page 2 of 2

	Yes	Partly	No

Does your organization's register of aspects and impacts identify the significant aspects and impacts arising from:

- All normal operating conditions?
- All normal activities, products and processes?
- All or any abnormal activities, products and processes?
- All accidents and potential emergency situations associated with your activities, products and processes?
- All past, present and planned activities, products and processes?
- The full life cycle of your organization's products?

Is your organization's register reviewed regularly and revised accordingly?

Is your organization's register documented and presented in a clear, concise and easy-to-understand format?

Does your organization's register differentiate between direct and indirect aspects and impacts?

Does your organization's register include the procedures employed to identify environmental aspects and impacts and their significance?

Does your organization's register identify the procedures or instructions employed to assess significant environmental aspects and impacts?

Is your organization's register kept in the environment management manual?

Chapter 4

The environmental policy

Objective of the chapter

The objective of this chapter is to explain what the environmental policy is. This chapter will provide you with the skills necessary to develop an environmental policy as the starting point for a functional EMS. After finishing this chapter, you should be able to answer the following questions:

✓ What is the environmental policy?

✓ Why is the environmental policy important?

✓ What is required of the environmental policy to meet the requirements of ISO 14001 and EMAS?

✓ What written procedures are required for the environmental policy?

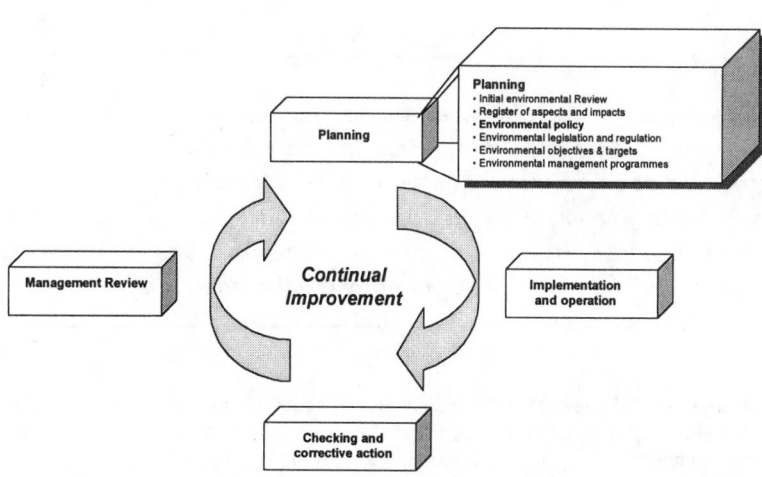

Introduction to the environmental policy

A policy is a set of rules or principles (usually formal and documented) that an individual, company or organization adopts for a chosen course of action. The environmental policy is no different. It is *a formal and documented set of principles and intentions with respect to the environment.* Essentially, the environmental policy is the guiding document for corporate environmental improvement and adherence to it is fundamental to the integrity and success of the entire EMS. The environmental policy should be developed after completing the initial environmental review.

In the case of your company, these overall principles and intentions would include continuous/continual environmental improvement and this would be met by minimizing the significant environmental impacts of your company's activities, products and processes.

Hint

While some might see the environmental policy simply as a public relations document posted on the wall in the company's reception, front lobby or coffee room, this is certainly not so. For an EMS to function properly and truly improve overall environmental performance, these principles and intentions must be respected. To achieve this you need both commitment from top management and a high level of awareness among all of the employees. As this is one of those areas that is easier said than done, you may want to pay particular attention to your environmental policy during your training programmes (see Chapter 10).

While there are slight variations between the ISO 14001 and EMAS definitions of environmental policy, the meaning and purpose in all cases is essentially the same. The environmental policy is a 'statement of your overall aims and principles of action with respect to the environment'[1] and is an essential requirement of a functional EMS.

Hint

When developing your policy, encourage participation and input from your employees and other interested parties. It cannot be stressed enough that your policy should be relevant to the activities, products and processes of your organization in order to convey the message that you have a sound understanding of your environmental aspects and impacts. Remember, the policy needs to be understood both by those within your organization and those external to it.

---

[1] Council Regulation (EEC) No 1836/93 of 29 June, 1993 allowing voluntary participation by companies in the industrial sector in a Community eco-management and audit scheme. *Official Journal of the European Communities*, 10 July, 1993 (OJ 10.7.93 L 168).

> **Certification tip**
>
> *As the guiding document for corporate environmental improvement your certifier will want to see that your policy has been developed in light of your identified significant environmental aspects and impacts as opposed to being 'cut and pasted' from elsewhere. As a rule of thumb, if your environmental policy could apply to another organization it probably needs improving. Be careful not to commit to courses of action that you know to be unrealistic and avoid making vague statements that do not imply commitment. Most importantly, you will want to prove that your policy is used as a guiding document. This can be done by showing that it is the basis for your environmental objectives and targets, and environmental management programmes, and that it is used as a benchmark for environmental performance.*

 Your environmental policy should be documented. This document should be considered a controlled document and be kept in your environmental management manual.

More precise requirements for an environmental policy are discussed in the next section, *Recommendations for successful EMS implementation*.

To develop your own policy you can use the blank and sample templates later in this chapter or on the floppy disk at the back of this book. For an example of what other companies have done visit the handbook web site at http://www.entropy-international.com/handbook/.

When you think you are finished, you can audit your performance and assess your success by using the check-list in the *Are you ready?* section at the end of this chapter.

Recommendations for successful EMS implementation

✓ Your environmental policy *must* state your organization's principles and intentions in relation to environmental performance.

✓ Your environmental policy *must* be appropriate to the nature, scale and significant environmental impacts of your company's activities, products and processes as identified in your initial environmental review and register of aspects and impacts.

✓ Your environmental policy *must* include a commitment to continual (continuous) improvement and the prevention of pollution based on a methodology such as the use of cleaner technology, BATNEEC (best available technology not entailing excessive costs) or EVABAT (economically viable application of the best available technology).

✓ Your environmental policy *must* include a commitment to comply with all environmental legislation and regulations associated with your organization's identified significant aspects and corresponding environmental impacts, and comply with other requirements to which your organization may subscribe, such as the ICC Principles for Environmental Management and the EMAS Good Management Practices.

✓ Your environmental policy *must* be documented, implemented, maintained and reviewed and be made known to all employees.

✓ Your environmental policy *must* be endorsed by senior management

✓ Your environmental policy *must* be made available to the public.

✓ Your environmental policy *must* provide for the setting, communicating and reviewing of your organization's objectives and targets.

✓ Your environmental policy *should* be written in clear, concise and non-technical language interpretable by both internal and external parties.

✓ Your environmental policy *should* include a commitment to the development and implementation of an environmental management system in your organization.

✓ Your environmental policy *should* include a commitment to the development of, and adherence to, corporate standards in the absence of legislation.

✓ Your environmental policy *should* embody a life cycle approach to the impact of your organization's activities, products and processes.

Dailuaine's Environmental Policy

Sample template

✓ Dailuaine is committed to the protection and enhancement of the environment. Improving our overall corporate environmental performance in all our operations is a major and continuing priority to be achieved by implementing and maintaining an environmental management system, and adhering to this environmental policy.

✓ Dailuaine endeavours to minimize the environmental impact of all of its activities, products and processes throughout their life cycle as identified in our initial environmental review. In particular we aim to minimize our emissions to air, reduce water use and reduce effluent to water by applying the most economically viable application of the best available technology and by adopting the principle of pollution prevention.

✓ To achieve such corporate environmental improvement, Dailuaine has set, and will maintain, review and revise, environmental objectives and targets with the aim of continually improving our environmental performance.

✓ Dailuaine commits to complying with all legislation, regulations and industry codes associated with our environmental impact and, where no legislation exists, we endeavour to set corporate standards to meet our overall objective of improved corporate environmental performance.

✓ Dailuaine will continually relate environmental considerations to wider commercial objectives and responsibilities to shareholders. In addition to this, we ensure that environmental issues and the views of interested parties, employees and the local community are taken into account in strategic decisions affecting the environment and that all new products, processes and investment proposals will be evaluated prior to approval to assess their likely environmental impacts.

✓ All employees of the company are expected to conduct their work in a manner compatible with the company's environmental policy and objectives. At Dailuaine, dialogue will be maintained to ensure that all employees are aware of the environmental policy and participate in environmental work at the site.

✓ Dailuaine conducts regular environmental reviews of all its operations, as stipulated by the environmental management system, to ensure compliance with the policy and aims to be an environmentally responsible organization by actively improving environmental performance in accordance with the ISO 14001 standard for environmental management.

Signed,

Grant Fromage

Grant Fromage,
Managing Director, Dailuaine

Are you ready?

EMS Checklist **Environmental Policy** *page 1 of 1*	■ Yes	◤ Partly	☐ No
Does your environmental policy state your organization's principles and intentions in relation to its overall environmental performance?	☐	☐	☐
Is your environmental policy appropriate to the nature, scale and environmental impacts of your organization's activities, products and processes?	☐	☐	☐
Does your environmental policy include a commitment to continual improvement and the prevention of pollution based on an acceptable methodology?	☐	☐	☐
Does your environmental policy include a commitment to comply with relevant environmental legislation, regulations and other requirements to which your organization subscribes?	☐	☐	☐
Is your environmental policy documented, implemented, maintained, reviewed and made known to all employees?	☐	☐	☐
Does top management endorse your environmental policy?	☐	☐	☐
Is your environmental policy available to the public?	☐	☐	☐
Does your environmental policy provide the framework for setting and reviewing environmental objectives and targets?	☐	☐	☐
Is your environmental policy clear, concise and written in non-technical language interpretable by both internal and external parties?	☐	☐	☐
Does your environmental policy include a commitment to the development and implementation of your EMS?	☐	☐	☐
Does your environmental policy include a commitment to the development of, and adherence to, corporate standards in the absence of legislation?	☐	☐	☐
Does your environmental policy embody a life cycle approach to the environmental impacts of your organization's activities, products and processes?	☐	☐	☐

Chapter 5

The register of environmental legislation and regulations

Objective of the chapter

The objective of this chapter is to explain what the register of environmental legislation and regulations is. This chapter will provide you with the skills necessary to develop your own register of environmental legislation and regulations sufficient to maintain a functional EMS. After finishing this chapter, you should be able to answer the following questions:

✓ What is the register of environmental legislation and regulations?

✓ Why is the register of environmental legislation and regulations important?

✓ What must be included in the register of environmental legislation and regulations to meet the requirements of ISO 14001 and EMAS?

✓ What written procedures are required for the register of environmental legislation and regulations?

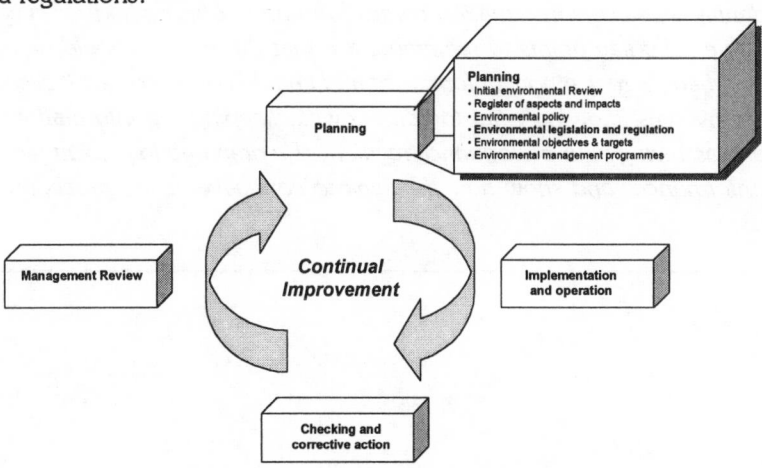

Introduction to environmental legislation and regulations

As mentioned in the previous chapter on environmental policy, one of the fundamental requirements of the environmental policy is that it contains a stated commitment to comply with all relevant environmental legislation and regulations by which the organization may be bound. Consequently, for an EMS to be truly effective (and certifiable) it must ensure that all its activities, products and processes are themselves compliant with relevant environmental legislation and regulations.

Just as you need to identify your organization's aspects and impacts to manage them effectively, to commit to legislative and regulatory compliance presupposes that your organization has identified and documented exactly what legislation and regulations it will comply with. Thus *a register of environmental legislation and regulations is a list of all the relevant environmental legislation and regulations by which your organization is bound.*

Hint

Due to the level of precision needed to assure compliance with environmental legislation and regulations it can be helpful to take advantage of proprietary services that offer concise reviews of the relevant requirements. Links to such services can be found on the handbook web site, http://www.entropy-international.com/handbook/.

Certification tip

As this is a major requirement of an EMS, your certifier will want to know that you have addressed all of the relevant legislation and regulations applicable to your site and that the system ensures compliance. If you use an external service, such as a commercial software package or database service provider, be sure you can identify where the information comes from and your method for keeping it up to date. The key points to remember are that the system should be able to identify when there is a nonconformance and enable you to correct it accordingly. This involves knowing what to comply with, keeping the information up to date and assessing your performance regularly. Be prepared to present records of nonconformances and show how they link to corrective action procedures and records.

📄 You must have a written procedure for identifying and having continued access to relevant legislation, regulations and other requirements to which your organization subscribes (see Chapter 9).

📄 You must have a written procedure for periodically reviewing and evaluating compliance with relevant legislation, regulations and other requirements to which your organization subscribes (see Chapter 9).

📑 Your register of environmental legislation and regulations should be documented. These documents should be considered controlled documents and be kept in your environmental management manual.

As with environmental policy and the register of environmental aspects and impacts, the definitions of and requirements for a register of environmental legislation and regulations differ slightly between ISO 14001 and EMAS. However, in both cases the register of environmental legislation and regulations is effectively the same thing and is included in each of the EMS requirements for the same reason.

In ISO 14001, the organization must *identify* what legislation and regulations it *subscribes* to, while in EMAS you must *record all legislative and regulatory requirements* associated with the activities, products and processes of the company. In short, your company must maintain an up-to-date list of all environmental legislative and regulatory requirements pertaining to any of your activities, functions, products or processes that have a significant environmental impact.

More precise requirements for a certifiable register of environmental legislation and regulations are discussed in the next section, *Recommendations for successful EMS implementation*.

💾 To develop your own register you can use the blank and sample templates later in this chapter or on the floppy disk at the back of this book.

When you think you are finished, you can audit your performance and assess your success by using the check-list in the *Are you ready?* section at the end of this chapter.

Recommendations for successful EMS implementation

✓ Your company *must* have a written procedure for identifying and having access to legal requirements and other regulations which are applicable to your organization, such as industry codes of conduct or any agreements with your authorities.

✓ Your organization *must* comply (or at very least commit to comply) with the identified environmental legislation and regulations associated with your organization's identified significant aspects and impacts and *must* comply with other requirements to which your organization subscribes, such as agreements with authorities or operational limits. (N.B. Commitment to comply is sufficient to meet the requirements of ISO 14001.)

✓ While it is *not* an explicit requirement of either ISO 14001 or EMAS, your organization *should* compile a register of all environmental legislation and regulations associated with your organization's identified aspects and corresponding significant environmental impacts (as identified in your initial environmental review). This should include other requirements to which your organization subscribes, such as agreements with local authorities or industry codes of conduct.

✓ The register of environmental legislation and regulations *must* be kept up to date and regularly reviewed to ensure that your organization complies with the legislation and regulations identified.

✓ The register of environmental legislation and regulations *should* identify the regulatory bodies associated with the identified legislation and regulations and briefly explain their activities and jurisdiction.

✓ The register of environmental legislation and regulations *should* include reference to *all* the regulatory instruments within which your organization must operate. Such a register *would* include:

● Any authorization or permission required for your activities, functions, production or processes

● Any site planning agreements, permits etc.

● Any regulation covering emissions to air, such as quality and/or quantity restrictions

- Any regulation covering discharges to water, such as quality and/or quantity restrictions

- Any regulation covering water use

- Any regulation covering waste disposal, such as due care, packaging use and disposal or minimization strategies

- Any regulation covering the use, storage and disposal of hazardous or special substances

- Any regulation covering contaminated land or the potential for land contamination

- Any regulation covering energy use, fuel use, land use and natural resource use

- Any regulation covering discharges of thermal energy, noise, odour, dust, vibrations and visual impact

- Any regulation covering aspects of worker health and safety that may also have environmental applicability

- Any regulation covering the protection of biodiversity or impact on the local or global ecological environment

✓ The register of environmental legislation and regulations *should* directly link identified legislation or regulation to your activities, products or processes.

Register of Legislation and Regulations	Blank template

Company Name:	Document Version:
Department/Site:	Issue/Revision Date:
Updated by:	Replaces Version:
Approved by:	Page of

Table of contents	Page

Register of Legislation and Regulations	Blank template

Company Name:	**Document Version:**
Department/Site:	**Issue/Revision Date:**
Updated by:	**Replaces Version:**
Approved by:	**Page of**

Section 1: Introduction

Section 2: Responsibility

Register of
Legislation and Regulations

Blank template

Company Name:

Department/Site:

Updated by:

Approved by:

Document Version:

Issue/Revision Date:

Replaces Version:

Page   of

Section 3: Waste legislation, regulations and authorizations

List of relevant legislation and regulations

Use similar
templates for
Sections 4-7

Register of **Legislation and Regulations**	*Blank template*

Company Name:	**Document Version:**
Department/Site:	**Issue/Revision Date:**
Updated by:	**Replaces Version:**
Approved by:	**Page of**

Title of legislation, regulations and authorizations
3.1

Description of legislation, regulation or authorization and associated authority, its activities and jurisdiction

Relevance to your organization

Associated activities, products or processes

Use similar templates for Sections 4-7

Register of Legislation and Regulations	*Blank template*

Company Name:	**Document Version:**
Department/Site:	**Issue/Revision Date:**
Updated by:	**Replaces Version:**
Approved by:	**Page of**

Section 8: Legislation, regulation and authorization matrix

Title of legislation, regulation or authorization	Activity, product or process affected

Register of **Legislation and Regulations**		*Sample template*

Company Name: United Distillers	**Document Version:** RLR001V1
Department/Site: Dailuaine Distillery	**Issue/Revision Date:** 31/06/97
Updated by: Wirral Green	**Replaces Version:** None
Approved by: Lotta Gutz	**Page** 1 **of** 60

Register of **Legislation and Regulations**	*Sample template*

Company Name: United Distillers   **Document Version:** RLR001V1

Department/Site: Dailuaine Distillery   **Issue/Revision Date:** 31/06/97

Updated by: Wirral Green   **Replaces Version:** None

Approved by: Lotta Gutz   **Page   2   of   60**

Section 1: Introduction

This document, the register of legislation and regulations, forms a crucial part of the environmental management system at Dailuaine distillery. This document is meant to identify all legislation, regulations or authorizations that are associated with any and all activities, products and/or processes at Dailuaine that have, or could have if uncontrolled, a significant environmental impact. Activities, products and processes that have, or could have if uncontrolled, significant environmental impacts are those as identified in the initial environmental review and register of aspects and impacts (RAI001) conducted in May of 1997. The findings of this review are compiled in the initial environmental review report that forms part of the environmental management manual.

This register shall be updated quarterly to ensure that Dailuaine distillery meets one of its main environmental policy objectives to comply regularly with all environmental regulations associated with its activities, products and processes and to ensure that Dailuaine continually improves its overall corporate environmental performance.

Section 2: Responsibility

It is the overall responsibility of top management at Dailuaine distillery to ensure that this register is kept up to date and that Dailuaine distillery complies with all environmental legislation and regulations associated with its activities, products and processes.

It is the individual responsibility of the Environmental Manager, Wirral Green, to update the register and amend it accordingly and inform relevant personnel of any changes that may affect them. This will be done quarterly, starting with the first quarter of each calendar year.

It shall be the responsibility of the Production Manager, Will Getitdon, to inform the Environmental Manager of any changes to operational practices, procedures or substances that may affect legislative, regulatory or policy compliance.

It shall be the responsibility of the VP for Production, Lotta Gutz, to ensure that appropriate resources are made available to accomplish the aforementioned tasks and responsibilities.

Register of Legislation and Regulations	*Sample template*

Company Name:	United Distillers	**Document Version:**	RLR001V1
Department/Site:	Dailuaine Distillery	**Issue/Revision Date:**	31/06/97
Updated by:	Wirral Green	**Replaces Version:**	None
Approved by:	Lotta Gutz	**Page** 3 **of** 60	

Section 3: Waste legislation, regulations and authorizations

List of waste legislation and regulations relevant to Dailuaine

3.1 Environmental Protection Act 1990 (EPA 90) – Part II

3.2 Control of Pollution (Amendment) Act 1989

3.3 Controlled Waste (Registration of Carriers and Seizure of Vehicles) Regulation 1991 (SI 1991/1624)

3.4 Controlled Waste Regulation 1992 (SI 1992/588), amended by SI 1993/566

3.5 Environmental Protection (Duty of Care Regulations 1991) SI 1991/2839

3.6 Special Waste Regulations 1996 (SI 1996/972)

3.7 Waste Management Licensing Regulations 1994 (SI 1994/1056), as amended by SI 1995/288 and SI 1995/1950

3.8 EU Regulation 259/93 on the Supervision and Control of Shipments of Waste within, into and out of the European Community

3.9 The Carriage of Dangerous Goods (Classification, Packaging and Labelling) and Use of Transportable Pressure Receptacles Regulations 1996 (CPL2)

3.10 The Carriage of Dangerous Goods by Road Regulations 1996 (CDG Road)

3.11 Environment Act 1995 – Part V

3.12 Landfill Tax Regulations 1996 (SI 1996/1527)

3.13 The Landfill Tax (Quantifying Material) Order 1996 (SI 1996/1528)

3.14 The Landfill Tax (Contaminated Land) Order 1996 (SI 1996/1529)

3.15 Draft Producers Responsibility Obligation (Packaging Waste) Regulations

Register of Legislation and Regulations	Sample template

Company Name:	United Distillers	**Document Version:**	RLR001V1
Department/Site:	Dailuaine Distillery	**Issue/Revision Date:**	31/06/97
Updated by:	Wirral Green	**Replaces Version:**	None
Approved by:	Lotta Gutz	**Page** 4 **of** 60	

Title of legislation, regulations and authorizations

3.1 Environmental Protection Act 1990 – Act of Parliament

Description of legislation, regulation or authorization and associated authority, its activities and jurisdiction

The Environmental Protection Act 1990 is an act of parliament of the United Kingdom. This national Act replaces the provisions of the Control of Pollution Act (COPA) of 1974 relating to waste on land. It imposes a duty of care on waste producers and makes further provisions in relation to controlled waste. Section 34 of this Act states that *it is the duty of any person who imports, produces, carries, keeps, trades, treats or disposes of controlled waste or as a broker has control of such waste, to take all measures applicable to him in that capacity as are reasonable to:*

- *Prevent the escape of waste from his control or that of any other person as defined in Part II of the Act,*

- *Ensure that the waste is transferred only to an authorized person and that the waste is accompanied by a written description of the waste sufficient to prevent other persons contravening these regulations.*

The Environmental Protection Act 1990 is under the jurisdiction of the Environment Agencies throughout the UK who have an overall responsibility for protecting and enhancing the environment in an integrated manner.

Relevance to your organization

- Dailuaine distillery must ensure that all employees involved with any waste from the site are aware of our duty of care as waste producers.

- Dailuaine distillery takes all measures possible to prevent the escape of waste from our control.

- Dailuaine distillery must ensure that only authorized persons transfer our waste and that all such transfers are accompanied by a written description.

Associated activities, products or processes

At Dailuaine distillery, the following processes are affected by the Environmental Protection Act 1990 – Act of Parliament:

Milling, Warehouse, Evaporation, Dark Grains Plant and Effluent Treatment

Register of Legislation and Regulations	Sample template

Company Name: United Distillers	**Document Version:** RLR001V1
Department/Site: Dailuaine Distillery	**Issue/Revision Date:** 31/06/97
Updated by: Wirral Green	**Replaces Version:** None
Approved by: Lotta Gutz	**Page** 58 **of** 60

Section 8: Legislation, regulation and authorization matrix

Title of legislation, regulation or authorization	Activity, product or process affected
1. The Environmental Protection Act 1990	Milling, Warehouse, Evaporation, Dark Grains Plant and Effluent Treatment.
2. The Environment Act 1995 – Act of Parliament	Milling, Mashing & Fermenting, Distilling, Warehouse, Evaporation, Dark Grains Plant and Effluent Treatment.
3. Special Waste Regulations 1996, SI 1996/972 – Statutory Instrument	Evaporation, Dark Grains Plant and Effluent Treatment.
4. Environmental Protection (Duty of Care) Regulations, SI 1991/2829 – Statutory Instrument	Milling, Mashing & Fermenting, Distilling, Warehouse, Evaporation, Dark Grains Plant and Effluent Treatment.
5. Chemicals Hazards and Packaging Regulations, SI 1993/1746 – Statutory Instrument	Evaporation and Dark Grains Plant
6. Waste Management and Licensing Regulation, SI 1994/1056 – Statutory Instrument	Milling, Warehouse, Evaporation, Dark Grains Plant and Effluent Treatment.
7. Controlled Waste (Amendment) Regulations, SI 1993/566 – Statutory Instrument	Warehouse, Evaporation, Dark Grains Plant and Effluent Treatment.
8. Collection and Disposal of Waste Regulations, SI 1988/819 – Statutory Instrument	Milling, Warehouse, Evaporation, Dark Grains Plant and Effluent Treatment.
9. The Control of Pollution (Special Waste) (Amendment) Regulation, SI 1988/1790 – Statutory Inst.	Evaporation, Dark Grains Plant and Effluent Treatment.
10. Decision 94/904/EC establishing a list of hazardous waste pursuant to 91/689/EEC	Evaporation, Dark Grains Plant and Effluent Treatment.
11. Directive 91/689/EEC on Hazardous Waste	Evaporation, Dark Grains Plant and Effluent Treatment.
12. Directive 75/439/EEC on Waste Oils	Milling, Mashing & Fermenting, Distilling, Warehouse, Evaporation, Dark Grains Plant and Effluent Treatment.
13. Directive 91/156/EEC on Waste	Milling, Warehouse, Evaporation, Dark Grains Plant and Effluent Treatment.
14. Directive 94/62/EC on Packaging and Packaging Waste.	Milling, Warehouse, Evaporation, Dark Grains Plant and Effluent Treatment.

Are you ready?

<table>
<tr><td colspan="4">

EMS Check-list
Environmental Legislation and Regulations

page 1 of 2

</td><td>■
Yes</td><td>◪
Partly</td><td>▨
No</td></tr>
<tr><td>Do you have a written procedure for identifying and having access to legal requirements and other regulations that are applicable to your organization?</td><td>❑</td><td>❑</td><td>❑</td></tr>
<tr><td>Do you comply with all the identified environmental legislation and regulations and with other requirements to which your organization subscribes?</td><td>❑</td><td>❑</td><td>❑</td></tr>
<tr><td>Have you compiled a register of all environmental legislation and regulations associated with your organization's identified aspects and corresponding significant environmental impacts, including other requirements to which your organization subscribes?</td><td>❑</td><td>❑</td><td>❑</td></tr>
<tr><td>Is your register of environmental legislation and regulations kept up to date and revised when necessary?</td><td>❑</td><td>❑</td><td>❑</td></tr>
<tr><td>Does your register identify the regulatory bodies associated with the identified legislation and regulations and briefly explain their activities and jurisdiction?</td><td>❑</td><td>❑</td><td>❑</td></tr>
<tr><td>Does your register include reference to authorizations or permits required by your organization?</td><td>❑</td><td>❑</td><td>❑</td></tr>
<tr><td>Does your register include all site planning arrangements required for the site?</td><td>❑</td><td>❑</td><td>❑</td></tr>
<tr><td>Does your register include all relevant regulations covering air emissions including quality and/or quantity restrictions?</td><td>❑</td><td>❑</td><td>❑</td></tr>
<tr><td>Does your register include all relevant regulations covering discharges to water including quality and/or quantity restrictions?</td><td>❑</td><td>❑</td><td>❑</td></tr>
<tr><td>Does your register include all relevant regulations covering water use?</td><td>❑</td><td>❑</td><td>❑</td></tr>
</table>

EMS Check-list
Environmental Legislation and Regulations

page 2 of 2

	■	◧	▢
	Yes	Partly	No

Does your register include all relevant regulations covering waste disposal such as due care, packaging use, and disposal or minimization strategies? ☐ ☐ ☐

Does your register include all relevant regulations covering the use, storage and disposal of hazardous or special substances? ☐ ☐ ☐

Does your register include all relevant regulations covering contaminated land or the potential for land contamination? ☐ ☐ ☐

Does your register include all relevant regulations covering energy use, fuel use, land use and natural resource use? ☐ ☐ ☐

Does your register include all relevant regulations covering discharges of thermal energy, noise, odour, dust, vibrations and visual impact? ☐ ☐ ☐

Does your register include all relevant regulations covering aspects of worker health and safety that may also have environmental applicability? ☐ ☐ ☐

Does your register include all relevant regulations covering the protection of biodiversity or impact on the local or global ecological environment? ☐ ☐ ☐

Does your register directly link identified legislation or regulation to individual activities, products or processes or your organization's activities? ☐ ☐ ☐

Chapter 6

Environmental objectives and targets

Objective of the chapter

The objective of this chapter is to explain what environmental objectives and targets are. This chapter will provide you with the skills necessary to develop your own environmental objectives and targets sufficient to maintain a functional EMS. After finishing this chapter, you should be able to answer the following questions:

✓ What are environmental objectives and targets?

✓ Why are environmental objectives and targets important?

✓ What is required of environmental objectives and targets to meet the requirements of ISO 14001 and EMAS?

✓ What written procedures are required for environmental objectives and targets?

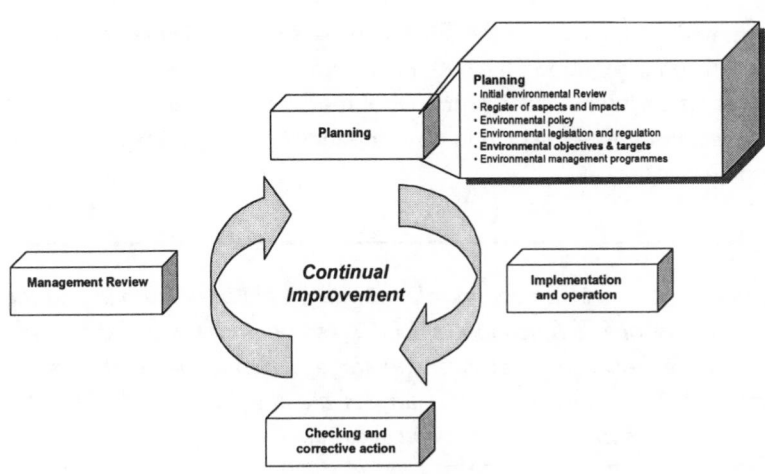

Planning
- Initial environmental Review
- Register of aspects and impacts
- Environmental policy
- Environmental legislation and regulation
- **Environmental objectives & targets**
- Environmental management programmes

Planning

Management Review

Continual Improvement

Implementation and operation

Checking and corrective action

Introduction to objectives and targets

Having completed an initial environmental review, a register of both aspects/impacts and legislation/regulation, and an environmental policy, you are well on your way to developing a functional and certifiable EMS for your organization.

At this point, you have completed two fundamental tasks. First, by completing the initial environmental review and preparing the registers you have identified and documented what your organization's impact on the environment is. Second, by preparing an environmental policy you have clearly established, in writing, your intention to improve overall environmental performance by minimizing the significant environmental impacts identified in the initial environmental review and documented in the register of environmental aspects and impacts.

By now, therefore, you have established, in writing, your willingness to improve environmental performance (environmental policy) and what it is that needs to be improved (IER/register of environmental aspects and impacts).

Your next step is to determine what will actually be done to achieve this improvement. Objectives and targets are precisely this: an identification of what will be done to improve.

Environmental objectives are the broad goals that your organization sets in order to improve environmental performance. Environmental objectives are goals such as 'reduce water use' or 'improve energy efficiency'.

Environmental targets are set performance measurements that must be met to realize a given objective. Targets are measurable and quantifiable statements, such as 'By 10 cubic meters/day' or '50% within two years'. All environmental objectives must have at least one target (usually more) and all targets must relate directly to a stated objective.

As you can see in Figure 6.1 the EMS can be seen as a series of interconnected events, each contingent upon the successful completion of the event that precedes it. Meeting the stated intentions of your environmental policy depends entirely on meeting the objectives you set, which in turn depends entirely on meeting the targets set for that objective.

Hint

Objectives and targets are commonly misinterpreted as being one and the same thing. In developing a functional EMS it is essential that you clearly understand the difference between objectives and targets. Remember that objectives are broad goals for improvement and targets are the benchmarks by which you measure progress in meeting the objectives set.

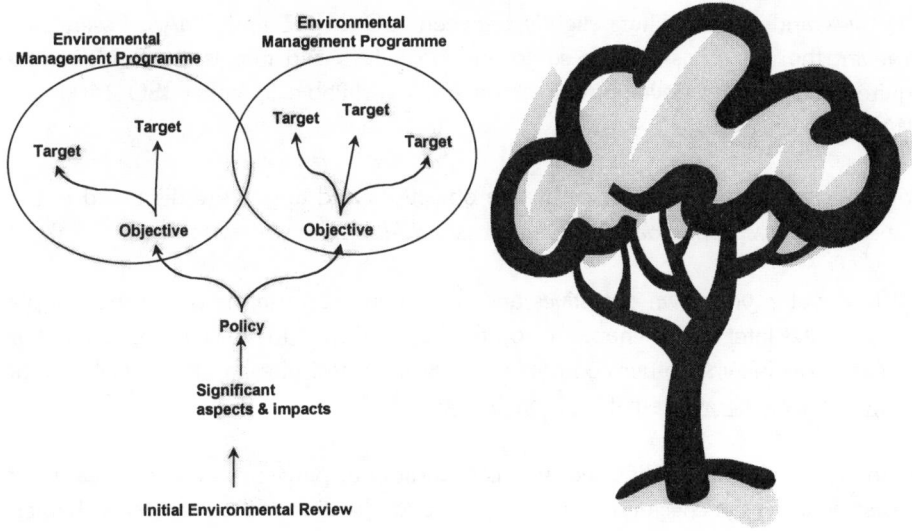

Figure 6.1 *The link between policy, objectives and targets*

Certification tip

It is necessary to be able to show how your significant environmental aspects and impacts link to your objectives. Similarly, it is necessary to show how objectives each link to one or more targets. It is useful to be able to show documented progress towards meeting your objectives and targets.

Your objectives and targets should be documented. These documents should be considered controlled documents and be kept in your environmental management manual.

As you will notice in the United Distillers example, the first and second environmental aspects of significance were the risk of explosion in the Dark Grains Plant and water use in the Mashing & Fermentation process. As the objective 'to reduce risk' is difficult to quantify and success at meeting such an objective is less tangibly measured, the second aspect (water use) will be used as an example. Therefore, in this case, an objective could be 'to reduce water use in the Mashing & Fermentation process' and a subsequent target for that objective could be 'by 50% by the year 2000'.

As with the components an EMS previously discussed, the exact definition of both objectives and targets differs slightly between ISO 14001 and EMAS. Essentially, however, the meanings are the same and objectives and targets are fundamental requirements for the development of an EMS certifiable to either ISO 14001 or EMAS.

More precise requirements for certifiable objectives and targets are discussed in the next section, *Recommendations for successful EMS implementation*.

To develop your own objectives and targets you can use the blank and sample templates later in this chapter or on the floppy disk at the back of this book. For examples of what other companies have done, visit the handbook web site at http://www.entropy-international.com/handbook/.

When you think you are finished, you can audit your performance and assess your success by using the check-list in the *Are you ready?* section at the end of this chapter.

Recommendations for successful EMS implementation

✓ Objectives are overall performance goals that *must* be reflected in the environmental policy.

✓ Objectives are goals that *must* be set by your organization.

✓ Objectives *must* be specific, realistic and achievable and aimed at the continual (continuous) improvement of your environmental performance.

✓ Objectives *must* be directly related to the significant environmental impacts of your activities, products and processes as determined by the findings of your IER.

✓ Objectives *must* be established, maintained and documented for all relevant activities, products and processes within your organization.

✓ Objectives *must* be consistent with the legislative and regulatory compliance requirements of your organization.

✓ Objectives *must* be set considering the views of internal and external stakeholders as well as financial, operational and organizational parameters of your organization.

✓ Objectives *must* be implemented, regularly reviewed and revised, as appropriate, with the endorsement of top management.

✓ Objectives *must* be supported with sufficient human and financial resources required for their achievement.

✓ Targets are detailed and quantified performance requirements developed to meet the environmental objectives you have set.

✓ Targets *must* be measurable and have set dates against which progress can be measured.

✓ Objectives and targets *should* encompass a preventative approach to pollution wherever practicable.

✓ Objectives and targets *should* make use of methodologies such as the use of cleaner technology, BATNEEC (best available technology not entailing excessive costs) or EVABAT (economically viable application of the best available technology), wherever possible.

✓ Objectives and targets *must* be set out in environmental management programmes (environmental action plans) specifying the action steps to be taken, schedules, resources and responsibilities required to meet the stated objective by the stated target date and target quantity/quality.

Environmental Objectives and Targets

Company Name:	**Document Version:**
Department/Site:	**Issue/Revision Date:**
Updated by:	**Replaces Version:**
Approved by:	**Page of**

Others involved:	**Objective No.** **Related Prog. No.**

Description of objective

Targets of stated objective	Estimated completion	Actual completion
1.	1.	1.
2.	2.	2.
3.	3.	3.

Evaluation procedure for objective and its targets

Environmental Objectives and Targets

Company Name:	United Distillers	**Document Version:**	OAT001V1
Department/Site:	Dailuaine Distillery	**Issue/Revision Date:**	01/07/97
Updated by:	Will Getitdon	**Replaces Version:**	None
Approved by:	G. Fromage	**Page** 1 **of** 1	

Others involved: Toby Dunn	**Objective No.**	OAT001
Ewan Meboth	**Related Prog. No.**	EMP001

Description of objective

To reduce water use in the Mashing & Fermentation process

Targets of stated objective	Estimated completion	Actual completion
1. To measure present water use in the Mashing & Fermentation process.	**1.** 15/07/97	**1.** 15/07/97
2. To develop water use reduction plan and operational procedures to ensure operations are carried out according to the plan developed.	**2.** 31/07/97	**2.** 28/07/97
3. To train all personnel in the Mashing & Fermentation process about the impact of water use, the reduction plan developed, the operational procedures to follow and how their activities affect this environmental aspect.	**3.** 31/08/97	**3.** 31/08/97

Evaluation procedure for objective and its targets

Will Getitdon is to review each objective every three months and each target monthly. Revisions will be made by W.G. to objectives and targets where necessary.

Are you ready?

<table>
<tr><td colspan="4">

EMS Check-list
Environmental Objectives and Targets
page 1 of 2
</td><td>■</td><td>◪</td><td>□</td></tr>
<tr><td colspan="4"></td><td>Yes</td><td>Partly</td><td>No</td></tr>
</table>

	Yes	Partly	No
Have site objectives and targets been set to improve your organization's environmental performance?	❏	❏	❏
Are your objectives overall performance goals?	❏	❏	❏
Are the objectives reflected in your environmental policy?	❏	❏	❏
Are your objectives detailed goals, in terms of environmental performance, which have been set at the site?	❏	❏	❏
Are your objectives specific, realistic and achievable?	❏	❏	❏
Are the objectives aimed at the continual (continuous) improvement of your organization's environmental performance?	❏	❏	❏
Are your objectives directly related to the significant environmental impacts of your activities, products or processes as determined by the findings of your IER?	❏	❏	❏
Have the objectives been established, maintained and documented for all relevant activities, products and processes within your organization?	❏	❏	❏
Are the objectives consistent with the legislative and regulatory compliance requirements of the site?	❏	❏	❏
Have the objectives been set considering the views of internal and external stakeholders as well as the financial, operational and organizational parameters of your organization?	❏	❏	❏
Have your objectives been implemented and are they regularly reviewed and revised, where necessary, with the endorsement of top management?	❏	❏	❏
Are your objectives supported with sufficient human and financial resources required for their achievement?	❏	❏	❏

EMS Check-list
Environmental Objectives & Targets

page 2 of 2

	Yes	Partly	No

	Yes	Partly	No
Are your targets detailed and quantifiable performance requirements?	☐	☐	☐
Have your targets been developed to meet the environmental objectives set?	☐	☐	☐
Are the targets measurable with set dates against which progress can be measured?	☐	☐	☐
Do your objectives and targets encompass a preventative approach to pollution wherever practicable?	☐	☐	☐
Do your objectives and targets make use of methodologies such as the use of cleaner technology, BATNEEC or EVABAT, wherever possible?	☐	☐	☐
Are your objectives and targets set out in an environmental management programme (action plan) specifying action steps to be taken, schedules, resources and responsibilities?	☐	☐	☐

Chapter 7

Environmental management programmes

Objective of the chapter

The objective of this chapter is to explain what environmental management programmes are. This chapter will provide you with the skills necessary to develop your own environmental management programmes sufficient to maintain a functional EMS. After finishing this chapter, you should be able to answer the following questions:

✓ What are environmental management programmes?

✓ Why are environmental management programmes important?

✓ What is required of environmental management programmes to meet the requirements of ISO 14001 and EMAS?

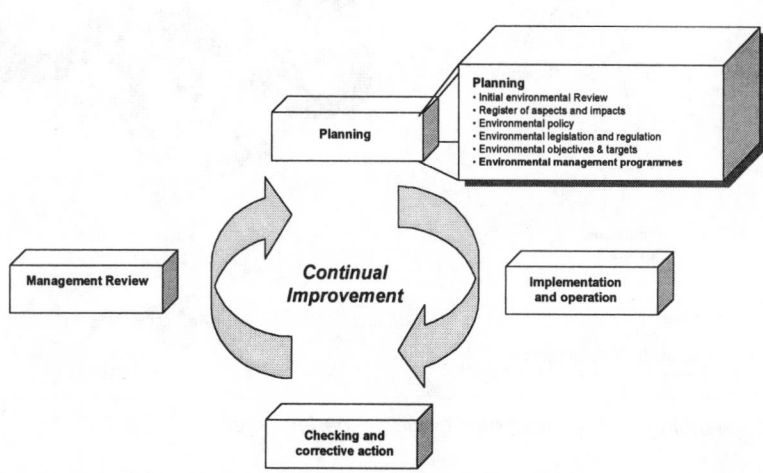

Introduction to the environmental management programme

Up to this point, you have determined:

✓   What is happening (IER)

✓   Your need to improve (significant aspects and impacts)

✓   Your intention to improve (policy)

✓   What improvement would entail (objectives and targets)

It is now essential to determine exactly what actions are required to improve. These actions are known as *environmental management programmes* and are essentially a detailed recipe for meeting the objectives and targets set. Logically, if a target is met, its correlating objective will similarly be met and the environmental policy will meet its stated intention, as depicted in Figure 7.1 below. Thus, for a given objective *the environmental management programme identifies how targets will be met, who is responsible for each of the activities required to meet that target and when those activities will be completed.*

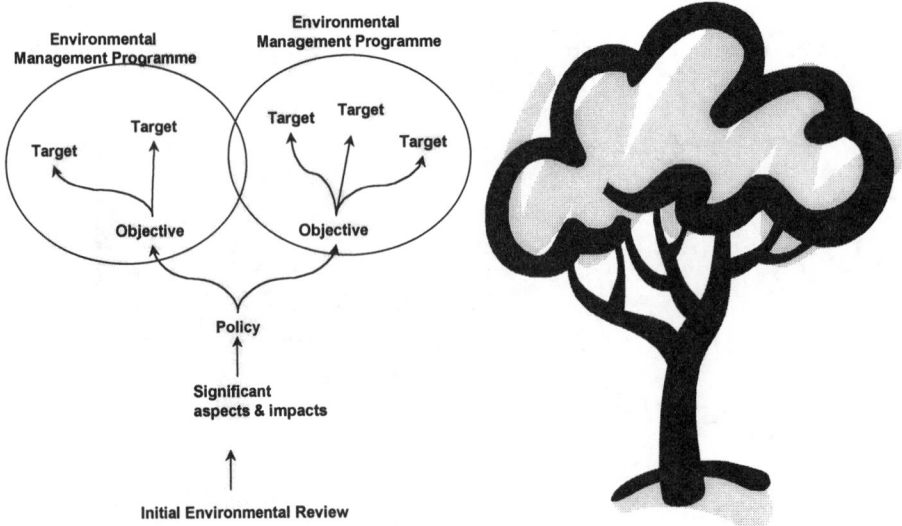

Figure 7.1 *Planning for improved environmental performance*

In the United Distillers example from Chapter 6, the objective 'to reduce water use in the Mashing and Fermentation process' by meeting the target of 'by 50% by the year 2000' would require actions if it were to be successful. An environmental management programme for such a target *could* be:

1   By July 15, 1997, Will Getitdon shall measure all present water use in the process. This will be done by evaluating water use for all areas of the process including cleaning.

2   By July 31, 1997, the Will Getitdon shall present the Managing Director with a plan that assesses the possibilities for water use reduction. The plan shall include an evaluation of process water use reduction, water recycling, effluent minimization, cleaner technologies and substitution technologies. This report will include an assessment of all training, resources, new equipment and time necessary to implement the recommendations of the plan.

3   By July 31, 1997, Will Getitdon shall develop procedures for all operations associated with water use and ensure that all personnel in the Mashing and Fermentation process are trained about water use, the operational procedures to follow and how their activities affect water use.

Although this management programme is very general, it does provide an example of some of the actions that would be required to meet the objective and target set.

Note

Your environmental management programmes are the vehicles by which your objectives and targets are realized. Successful implementation, control and revision of your programmes will fuel the engine of improved environmental performance!

Certification tip

For a smoother assessment, make sure your environmental management programmes identify resources and responsibilities required for achieving your stated objectives and targets. Be sure to develop environmental management programmes that address your significant environmental aspects and impacts in order of priority!

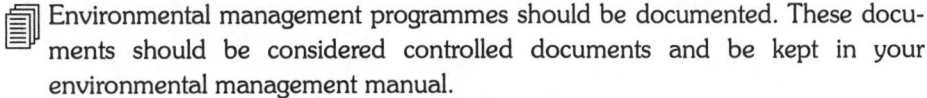 Environmental management programmes should be documented. These documents should be considered controlled documents and be kept in your environmental management manual.

As with the components an EMS previously discussed, the exact definition of 'environmental management programme' differs slightly between ISO 14001 and EMAS.

In EMAS, this set of actions is referred to as an *environmental programme*, while in ISO 14001 it is referred to as an *environmental management programme*. In both cases however, environmental management programmes are essentially the same thing. Environmental management programmes are required components of all certifiable EMSs and are the activities, measures, responsibilities and time-frames required to achieve your stated objectives.

More precise requirements for certifiable environmental management programmes are discussed in the next section, *Recommendations for successful EMS implementation*.

To develop your own environmental management programmes you can use the blank and sample templates later in this chapter or on the floppy disk at the back of this book. For examples of what other companies have done, visit the handbook web site at http://www.entropy-international.com/handbook/.

When you think you are finished, you can audit your performance and assess your success by using the check-list in the *Are you ready?* section at the end of this chapter.

Recommendations for successful EMS implementation

✓ You *must* establish and maintain environmental management programmes (environmental action plans) to meet the objectives and targets you have set.

✓ Your environmental management programmes are the recipes for achieving the objectives and targets that have been set to realize the environmental policy and thus improve overall corporate environmental performance.

✓ Your environmental management programmes *must* have milestones, deadlines (time-frames) and assigned responsibilities at every function and level of your company.

✓ Your environmental management programmes *must* establish the means and schedules for achieving objectives and targets and therefore should answer what, when, who, how and what next?

✓ Your environmental management programmes *must* be regularly reviewed and targets, budgets, responsibilities etc. *must* be revised accordingly.

✓ Your environmental management programmes *must* be revised in light of any new activities, products or processes.

✓ Environmental management programmes *must* be documented and *should* be kept in the environmental management manual.

✓ Environmental management programmes *should* be developed by those most closely associated with them – i.e. the department involved – but should be approved by, and be the ultimate responsibility of, upper management.

✓ The actions in your environmental management programmes *should* have their own correlating objective, description, budget, evaluation procedure, start and finish dates, and training requirements for those involved (where training may be needed).

✓ Evaluations *should* include how your programmes will be monitored, by whom, how problems and departures from the programmes will be dealt with and who is responsible for initiating and monitoring corrective action taken.

Environmental Management Programme	*Blank template*

Company Name:	Document Version:
Department/Site:	Issue/Revision Date:
Updated by:	Replaces Version:
Approved by:	Page of

Programme Title:

Others involved:	Programme No.
	Related Objective No.
	Budget Allocation:

Objective of programme

Description of programme

Targets of programme	Person responsible and target deadlines
1.	1.
2.	2.
3.	3.
4.	4.
5.	5.

Evaluation procedure

Training requirements

Programme start date:	Programme deadline:

Environmental Management Programme	*Sample template*

Company Name:	United Distillers	**Document Version:**	EMP001V1
Department/Site:	Dailuaine Distillery	**Issue/Revision Date:**	01/07/97
Updated by:	Wirral Green	**Replaces Version:**	None
Approved by:	G. Fromage	**Page** 1 **of** 1	

Programme Title: Water use reduction programme (Mashing & Fermentation process)

Others involved:	
	Programme No. EMP001
	Related Objective No. OAT001
	Budget Allocation: £ 25,500.00

Objective of programme

This programme has been established to meet objective no. OBJ001 to reduce water use in the Mashing and Fermentation process by 50% by the year 2000.

Description of programme

By July 15, 1997, Will Getitdon shall measure all present water use in the process. This will be done by evaluating water use for parts of the process including cleaning.

By July 21, 1997, Will Getitdon shall present the Managing Director with a plan that assesses the possibilities for water use reduction. The plan shall include an evaluation of process water use reduction, water recycling, effluent minimization, cleaner technologies and substitution technologies. This report will include an assessment of all training, resources, new equipment and time necessary to implement the recommendations of the plan.

By July 31, 1997, Will Getitdon shall develop procedures for all operations associated with water use and ensure that all personnel in the Mashing and Fermentation process are trained about water use, the operational procedures to follow and how their activities affect water use.

Targets of programme	**Person responsible and target deadlines**
1. Water use measurement	**1.** Will Getitdon, 15/07/97
2. Water use reduction plan	**2.** Will Getitdon, 21/07/97
3. Training needs assessment	**3.** Will Getitdon, 21/07/97
4. Develop operational procedures	**4.** Wirral Green, 31/07/97
5. Provide training	**5.** Will Getitdon, 31/07/97

Evaluation procedure

Wirral Green will regularly evaluate all targets weekly. Nonconformance with the programme will be corrected accordingly with all affected personnel to be notified.

Training requirements

Training on operational procedures and overall water use; training on good housekeeping methods; and training on technical options to be implemented if necessary.

Programme start date: 01/02/97   **Programme deadline:** 31/07/97

Are you ready?

EMS Check-list	■ Yes	◪ Partly	◻ No
Environmental Management Programme *page 1 of 1*			
Have you established and do you maintain environmental management programmes (environmental action plan) to meet your objectives and targets?	❑	❑	❑
Are your environmental management programmes the recipe for achieving your objectives and targets?	❑	❑	❑
Do your environmental management programmes have milestones, deadlines (time-frames) and assigned responsibilities?	❑	❑	❑
Do your environmental management programmes establish the means and schedules for achieving objectives and targets and answer what, when, who, how and what next?	❑	❑	❑
Are your environmental management programmes regularly reviewed (targets, budgets, responsibilities etc.) and revised accordingly?	❑	❑	❑
Are your environmental management programmes revised in light of any new activities, products or processes?	❑	❑	❑
Are your environmental management programmes documented and kept in the environmental management manual?	❑	❑	❑
Are your environmental management programmes the ultimate responsibility of upper management and have they been developed by those most closely associated with it?	❑	❑	❑
Do the actions in the environmental management programmes have their own correlating objective, description, budget, evaluation procedure, start and finish dates, and training requirements?	❑	❑	❑
Are the methods of evaluation for your environmental management programmes documented and agreed upon?	❑	❑	❑
Does this evaluation include how your programmes will be monitored, by whom, how problems and departures from your programmes will be dealt with and who is responsible for initiating and monitoring corrective action taken?	❑	❑	❑

Chapter 8

Structure and responsibilities

Objective of the chapter

The objective of this chapter is to explain what structure and responsibilities are. This chapter will provide you with the skills necessary to identify what structure and responsibility is required to maintain a functional EMS. After finishing this chapter, you should be able to answer the following questions:

✓ What are structure and responsibilities?

✓ Why are structure and responsibilities important?

✓ What level of structure and responsibilities are necessary to meet the requirements of ISO 14001 and EMAS?

✓ What written procedures are required for structure and responsibilities?

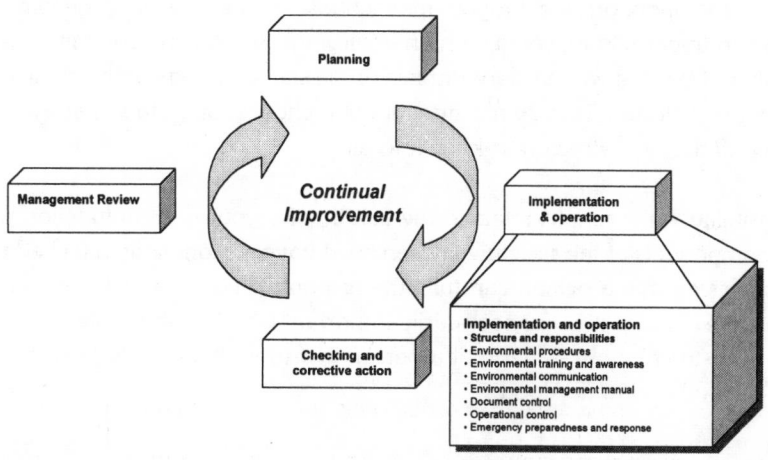

Introduction to structure and responsibilities

If you recall the 'Cycle of improvement' diagram introduced in Chapter 1, you will remember the parallel drawn to the development of an EMS. That diagram depicted the 'Stages of ISO 14001 implementation' and included planning, implementation, checking and corrective action, and management review. By now, you have completed:

- An environmental policy

- Register of environmental aspects and impacts

- Register of environmental legislation and regulation

- Environmental objectives and targets

- Environmental management programmes

Thus, the planning stage of the EMS is complete. At this point, you have developed the main building blocks of a functional EMS and it is now time to start putting these blocks together. This next step is the 'do' or 'implementation' stage of the cycle.

With the building blocks of your management system ready for assembly, to develop a functional EMS you now need to establish what form or 'structure' your system will have. Just as other management systems come in a variety of structures, the structure your EMS takes on will be unique to your company and reflect the organizational management that already exists.

The structure can be 'hierarchical' with one chief executive at the top and an increasing number of subordinates at each of the lower management levels it or it could be 'flatter' with fewer management levels and a greater number of participants at each level identified. You may decide to have an environmental manager in each department or operation on site or you may choose to have a single environmental manager in upper management with an environmental steering committee with representatives from the various departments of the site. Once again, the choice is yours and what is important is *not* what structure you choose, but rather that a structure *is* chosen and that that structure is known to all.

Responsibility, quite simply, refers to the fact that for your EMS to function properly, all the components of the structure chosen must have responsibility attached to them. Similarly, given that a person can fulfil the responsibilities of a position identified in the structure chosen, it is *not* necessarily important who is responsible for each position in that structure but rather that a person *is* responsible and that responsibility is clearly defined and understood.

In short, *structure refers to the administrative form of the EMS and responsibility refers to the roles, authorities and interrelations of the key personnel required to ensure the efficacy of the EMS and its chosen structure.*

Hint

The structure of your EMS should mirror that of, and fit into, existing management structures! It may help to draw the structure chosen and clearly allocate, by name and title, responsibility for all parts of the picture drawn. Many companies establish environmental steering committees and treat environmental tasks and responsibilities like individual projects with identified managers, budgets, resources required, completion dates etc.

Note

Top management must appoint a management representative who has identified responsibility for overall implementation and maintenance of your EMS. This manager must regularly report to top management about the EMS and its performance as a basis for management review and subsequent improvement to the EMS.

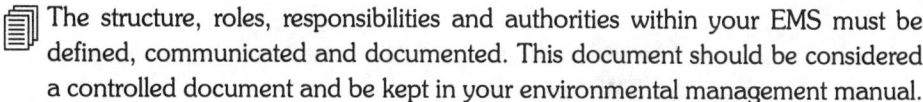 The structure, roles, responsibilities and authorities within your EMS must be defined, communicated and documented. This document should be considered a controlled document and be kept in your environmental management manual.

As with the main components of the EMS discussed in previous chapters, there are minor differences between ISO 14001 and EMAS for structure and responsibility. In EMAS for example, this concept of administrative structure and responsibility is referred to as *organization and personnel* while in ISO 14001 administrative structure and responsibility is referred to as *structure and responsibility*. In both cases however, the concept is the same. The EMS must have a structure and designated responsibility.

More precise requirements for structure and responsibility are discussed in the next section, *Recommendations for successful EMS implementation.*

To develop your own environmental structure and responsibility you may wish to refer to the sample templates later in this chapter. For examples of what other companies have done, visit the handbook web site at http://www.entropy-international.com/handbook/.

When you think you are finished, you can audit your performance and assess your success by using the check-list in the *Are you ready?* section at the end of this chapter.

Recommendations for successful EMS implementation

✓ All employees of your organization *must* be committed to the successful implementation and maintenance of your EMS.

✓ Top management *must* designate a management representative(s) and define their responsibility(ies) and authorities in relation to your EMS.

✓ All roles, responsibilities and authorities for implementing, operating and maintaining your EMS *must* be defined, documented and communicated within your organization.

✓ Top management *must* support these roles, responsibilities and authorities by providing appropriate human, financial, training and technical resources required by the people identified to meet their responsibilities in implementing, operating and maintaining your EMS.

✓ All roles, responsibilities and authorities *must* be approved and endorsed by top management.

✓ Roles, responsibilities and authorities *must* be established to ensure that all requirements of your EMS are established, implemented and maintained.

✓ Roles, responsibilities and authorities *must* be established for reporting on the performance of your EMS to top management.

✓ Roles, responsibilities and authorities *should* follow existing hierarchical structures and *should* be integrated into the overall responsibilities of all company personnel within your organization.

✓ You *must* develop a structure for the implementation, operation and maintenance of your EMS. This structure should include an organizational chart allocating the responsibilities, job descriptions and lines of communication between the various levels and functions of the structure.

✓ Your EMS may best be maintained and operated by an environmental management steering committee made up of middle and lower management with some practical knowledge about the environmental impact of your organization's operations.

Environmental Structure and Responsibility		*Sample chart*
Company Name: United Distillers	**Document Version:**	ESR001V1
Department/Site: Dailuaine Distillery	**Issue/Revision Date:**	01/07/97
Updated by: Wirral Green	**Replaces Version:**	None
Approved by: G. Fromage	**Page** 1 **of** 3	

Corporate Structure and Responsibility

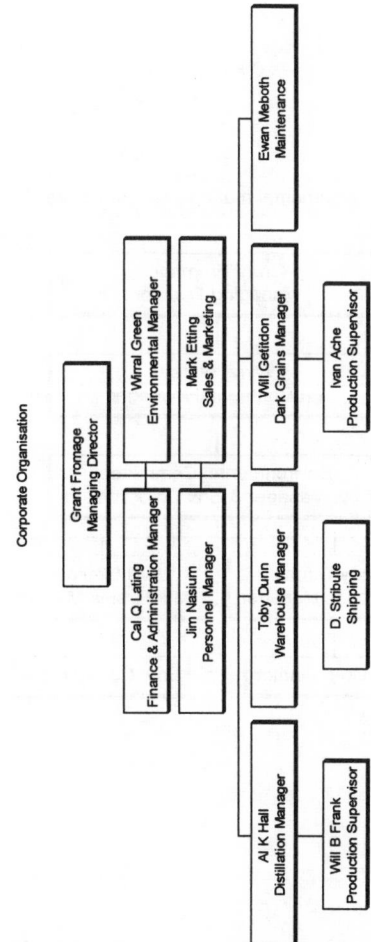

Environmental Structure and Responsibility	Sample chart

Company Name: United Distillers	**Document Version:** ESR001V1
Department/Site: Dailuaine Distillery	**Issue/Revision Date:** 01/07/97
Updated by: Wirral Green	**Replaces Version:** None
Approved by: G. Fromage	**Page** 2 **of** 3

Organizational Chart
with Environmental Responsibilities

Environmental management structure

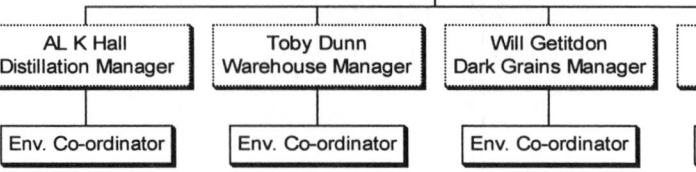

Environmental **Structure and Responsibility**		*Sample chart*
Company Name: United Distillers	**Document Version:**	ESR001V1
Department/Site: Dailuaine Distillery	**Issue/Revision Date:**	01/07/97
Updated by: Wirral Green	**Replaces Version:**	None
Approved by: G. Fromage	**Page** 3 **of** 3	

Organizational Chart for the
Environmental Management Steering Committee

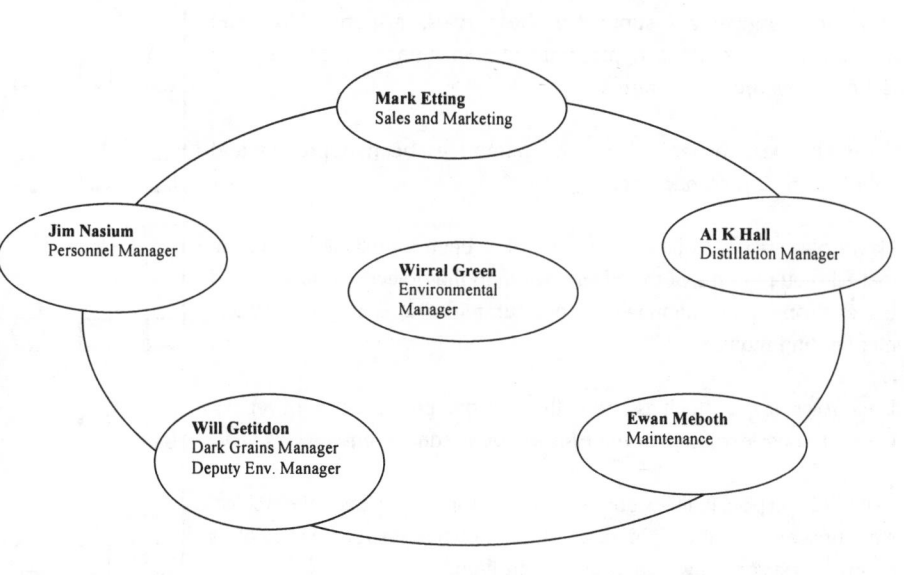

Are you ready?

EMS Check-list **Environmental Management** **Structure & Responsibilities** *page 1 of 1*	■ Yes	◪ Partly	☐ No
Are all the employees in your organization committed to the successful implementation and maintenance of your organization's EMS?	☐	☐	☐
Has top management designated specific management representative(s) and defined their responsibility(ies) and authorities in relation to the EMS?	☐	☐	☐
Are all roles, responsibilities and authorities for implementing, operating and maintaining your EMS defined, documented and communicated within your organization?	☐	☐	☐
Has top management supported these roles, responsibilities and authorities by providing appropriate human, financial, training and technical resources required?	☐	☐	☐
Have all roles, responsibilities and authorities been approved and endorsed by top management?	☐	☐	☐
Have roles, responsibilities and authorities been established to ensure that all requirements of the EMS (including the objectives, targets and the environmental management programme) are established, implemented and maintained?	☐	☐	☐
Have roles, responsibilities and authorities been established for reporting on the performance of the EMS to top management?	☐	☐	☐
Do roles, responsibilities and authorities follow existing hierarchical structures and are they integrated into the overall responsibilities of all company personnel within your organization?	☐	☐	☐
Have you developed a structure for the implementation, operation and maintenance of your EMS?	☐	☐	☐
Is this structure depicted by an organizational chart showing the responsibilities and the reporting lines between people identified?	☐	☐	☐

Chapter 9

Environmental procedures

Objective of the chapter

The objective of this chapter is to explain what environmental procedures are. This chapter will provide you with the skills necessary to develop your own procedures and identify what procedures are required to maintain a functional EMS. After finishing this chapter, you should be able to answer the following questions:

✓ What are environmental procedures?

✓ Why are procedures important?

✓ What procedures are necessary to meet the requirements of ISO 14001 and EMAS?

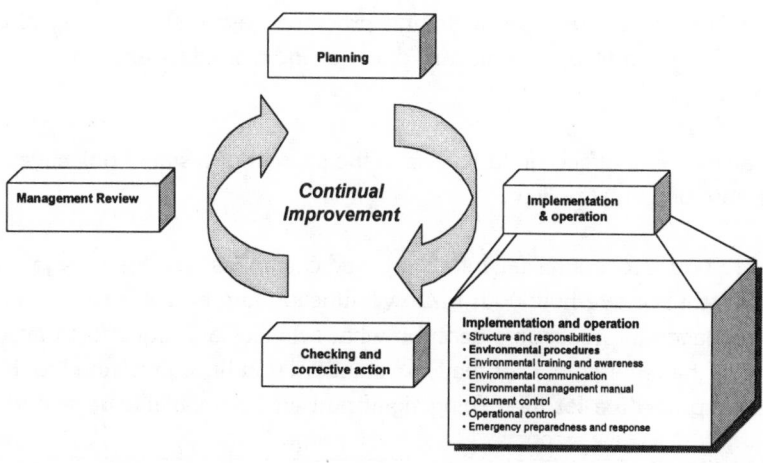

Introduction to environmental procedures

By now, your EMS should be taking shape. At this point you should have:

- Developed the guiding document for improved environmental performance (the environmental policy)

- Established what it is that needs to be improved (aspects and impacts)

- Established what improvement will entail (objectives, targets and management programmes)

- Established what management structure and responsibilities will be required to improve corporate environmental performance

Now it is time to develop procedures to ensure that your operations reflect the intentions and purpose of the emerging EMS. If you recall from Chapter 1, improved environmental performance is a direct result of minimizing significant environmental impacts by controlling the aspects of your operations that cause the impacts identified.

Environmental procedures are the step-by-step instructions that, if carried out properly, will control both your EMS and your organization's activities, products and processes (aspects). This will minimize correlating environmental impacts and thus improve the overall environmental performance of your organization.

While there are a number of procedures that a functional EMS must have, it is not always clear what actions would require a procedure. In short, procedures should be prepared for:

- All actions – activities, products and processes (aspects) – that have, or could have if uncontrolled, a significant direct or indirect effect (impact) on the environment

- All actions that affect, or could affect, the success of a stated objective and thus the environmental policy

Essentially, your procedures fall into two broad categories. The first category includes procedures that dictate how your EMS will function and how it is to be maintained. These 'managerial' procedures describe what activities are required to ensure your EMS meets the requirements of the EMS standard you have prescribed to. Examples include the procedure for identifying significant environmental aspects and impacts and the procedure for identifying the environmental training needs in your organization. These 'managerial' procedures are discussed in the handbook chapters to which they relate and are determined by the EMS requirements you follow to develop your EMS.

The second category includes procedures that dictate how your operations are to be controlled to minimize the significant environmental impacts associated with those operations. These 'operational' procedures are determined by (and should parallel) the significant environmental aspects of your operations as identified in your initial environmental review.

Hint

The terms 'procedure' and 'work instruction' are very similar and are often used interchangeably. A procedure is a documented description of how certain tasks are to be performed to ensure that the environmental policy and objectives and targets are not compromised and guarantee that all significant environmental aspects identified are controlled properly. Work instructions are more detailed step-by-step directions on how to carry out a certain task. Work instructions define the work required in terms of who is to perform it, when it is to commence, when it is to be completed, etc. An instruction can be included as part of a procedure or be a separate document. If your organization requires a large number of instructions to control its operations sufficiently, it may be a good idea to keep the procedures short and refer to more detailed work instructions where necessary.

Hint

If a procedure is not being followed, ask yourself why, as it most likely needs to be rewritten. Remember, you should always document what you intend to do, and do exactly what you have documented.

Certification tip

As procedures are required throughout your EMS, this will often be an area of concern during your audit and certification processes. At the minimum be sure you have prepared procedures that are required for a functional EMS (see the following 'Recommendations' section) and regularly check the validity of your procedures; i.e. are you doing what you have documented?

All of your EMS procedures should be documented. These documents should be considered controlled documents and be kept in your environmental management manual.

More precise requirements for environmental procedures are discussed in the next section, *Recommendations for successful EMS implementation*.

To develop your own environmental procedures you can use the blank and sample templates later in this chapter or on the floppy disk at the back of this book. For examples of what other companies have done, visit the handbook web site at http://www.entropy-international.com/handbook/.

When you think you are finished, you can audit your performance and assess your success by using the check-list in the *Are you ready?* section at the end of this chapter.

Recommendations for successful EMS implementation

✓ Environmental procedures form the basis for implementing the environmental management programmes (environmental action plans) and link responsibilities to desired outputs.

✓ Environmental procedures *must* be developed for all activities, products and processes that have, or could have if uncontrolled, a significant direct or indirect impact on the environment as indicated in the findings of your initial environmental review.

✓ Environmental procedures *must* be appropriate to the nature, complexity and scale of the activity, product or process that they are intended to control.

✓ Environmental procedures *must* be documented and should be included in the environment management manual.

✓ Environmental procedures specify who is to carry out what task with step-by-step instructions for how tasks are to be completed.

✓ Environmental procedures *must* include directions for dealing with departure from those procedures.

✓ Your EMS *must* have the following procedures:

- A procedure for identifying the environmental aspects and impacts associated with your activities, products and processes (see Chapter 2, Section IV, *The significance test for identified aspects and impacts*).

- A procedure to cover situations where an environmental aspect or impact is not controlled by your EMS and thus compromises the realization of your organization's environmental policy and stated objectives and targets.

- Operational procedures for controlling all activities, products and processes that have or could have a significant impact on the environment (see Chapter 2).

- A procedure for approving planned activities, products and processes (see Chapter 2).

- A procedure for identifying and having continued access to relevant legislation, regulations and other requirements relevant to your organization (see Chapter 5).

- A procedure for periodically reviewing and evaluating compliance with relevant legislation, regulations and other requirements relevant to your organization (see Chapter 5).

- A procedure for: identifying the training needs required to implement, operate and maintain your EMS; training all personnel about the importance of conformance with the your environmental policy, objective and targets and with the overall requirements of your EMS; ensuring that all personnel are trained on the significant impacts of activities, products and processes (actual and possible) of their own work activities; and ensuring that all personnel are trained on the consequences of departure from operating procedures (see Chapter 10).

- A procedure addressing procurement, purchasing and contract activities, to ensure that suppliers and those acting on behalf of your company are aware of and comply with your company's environmental policy (see Chapter 10).

- A procedure for internal communication within your organization about your organization's environmental aspects, impacts and the EMS as a whole (see Chapter 11).

- A procedure for receiving, documenting and responding to external communication associated with your organization's environmental performance or the EMS as a whole (see Chapter 11).

- A procedure for controlling all related EMS documents to ensure that: they can be located; they are reviewed, revised and approved regularly; current versions of documents are kept and available where needed; and obsolete documents are removed or otherwise marked accordingly (see Chapter 13).

- A procedure for identifying the likelihood of accidents and emergencies and the minimization, control and mitigation of environmental impacts associated with such situations (see Chapter 15).

- A procedure for regularly monitoring and measuring the functions, activities, products or processes that have or could have a significant impact on the environment (see Chapter 16).

- A procedure for defining responsibility and assigning authority for investigating the EMS for nonconformance with the stated requirements and rectifying that nonconformance (see Chapter 17).

- A procedure for identifying, maintaining and controlling environmental records including monitoring and measuring results, training records, audit findings and review reports (see Chapter 18).

- A procedure for periodic audits of the EMS (see Chapter 19).

Environmental Procedures		Blank template
Company Name:	**Document Version:**	
Department/Site:	**Issue/Revision Date:**	
Updated by:	**Replaces Version:**	
Approved by:	**Page of**	

Procedure:

| **Others involved:** | **Procedure No.** |
| | **Related Programme No.** |

Procedure purpose

Scope of procedure

Description of tasks involved with this procedure including definitions where necessary (1, 2, 3 etc.)

Expected results and actions in light of departure from this procedure

Related documentation:

Date procedure is to be reviewed (and revised where necessary):

Person responsible for updating this procedure:

<table>
<tr><td colspan="2"></td><td>Sample template</td></tr>
</table>

	Environmental Procedures

Company Name: United Distillers   **Document Version:** PROC001V1

Department/Site: Dailuaine Distillery   **Issue/Revision Date:** 16/07/97

Updated by: Will Getitdon   **Replaces Version:** None

Approved by: Wirral Green   **Page** 1 **of** 22

Procedure: Mashing and Fermentation cleaning procedure

Others involved: Toby Dunn   **Procedure No.** PROC001

Ewan Meboth   **Related Prog. No.** EMP001

Procedure purpose

The purpose of this procedure is to:

- ensure that cleaning in the Mashing and Fermentation process is conducted according to criteria that reduce water use.

Scope of procedure

These procedures will cover all wet cleaning in the Mashing and Fermentation process and will be followed at all times without exception.

Description of tasks involved with this procedure including definitions where necessary (1, 2, 3 etc.)

1. Before cleaning with water, area to be cleaned is to be swept (where possible) to ensure larger residuals are removed without the need for water.
2. Wash water is to be saved and used as the primary wash in the next cleaning process (backflow washing).
3. All water use in the process shall be measured and recorded.

Expected results and actions in light of departure from this procedure

It is expected that by 31/08/97 all wet cleaning in this process shall be consistent with this procedure. Identified departure from this procedure is to be evaluated by Ewan Meboth with corrective action documented and provided to those personnel involved and Wirral Green.

Related documentation: OAT001

Date procedure is to be reviewed (and revised where necessary):
01/09/97

Person responsible for updating this procedure: Ewan Meboth

Are you ready?

EMS Check-list Environmental Procedures *page 1 of 2*	■ Yes	◪ Partly	☐ No
Do your environmental procedures form the basis for implementing the environmental management programmes (environmental action plans) and link responsibilities to desired outputs?	☐	☐	☐
Have procedures been developed for all activities, products and processes that have, or could have if uncontrolled, a significant direct or indirect impact on the environment?	☐	☐	☐
Are your procedures appropriate to the nature, complexity and scale of the activity, product or process that they are intended to control?	☐	☐	☐
Are your procedures documented and included in the environment management manual?	☐	☐	☐
Do your procedures specify who is to carry out what task with step-by-step instructions for how tasks are to be completed?	☐	☐	☐
Do your procedures include directions for dealing with departure from those procedures?	☐	☐	☐
Do you have a procedure for identifying and controlling the environmental aspects and impacts associated with activities, products and processes?	☐	☐	☐
Do you have a procedure for performance criteria of all activities, products and processes that have or could have a significant impact on the environment?	☐	☐	☐
Is there a procedure for identifying and having continued access to relevant legislation, regulations and other requirements relevant to your organization?	☐	☐	☐
Is there a procedure for periodically reviewing and evaluating compliance with relevant legislation, regulations and other requirements relevant to your organization?	☐	☐	☐
Do you have a procedure for identifying the training needs required to implement, operate and maintain the EMS?	☐	☐	☐
Do you have a procedure for training all personnel about the importance of conformance with the environmental policy, objectives and targets and with the overall requirements of the EMS at the site?	☐	☐	☐
Do you have a procedure for training personnel about the significant impacts of activities, products and processes (actual and possible) of their own work activities?	☐	☐	☐

EMS Check-list
Environmental Procedures

page 2 of 2

	Yes	Partly	No

Do you have a procedure for training personnel in the consequences of departure from operating procedures?

☐ ☐ ☐

Do you have a procedure for identifying the likelihood of accidents and emergencies and the minimization, control and mitigation of environmental impacts associated with such situations?

☐ ☐ ☐

Do you have a procedure for internal communication about your organization's environmental aspects, impacts and the EMS as a whole?

☐ ☐ ☐

Do you have a procedure for receiving, recording and responding to external communication associated with your organization's environmental performance or the EMS as a whole?

☐ ☐ ☐

Do you have a procedure for regularly monitoring and measuring the activities, products or processes that have or could have a significant impact on the environment?

☐ ☐ ☐

Do you have a procedure for dealing with procurement and contract activities to ensure that suppliers and those acting on behalf of your organization are aware of and comply with your environmental policy?

☐ ☐ ☐

Do you have a procedure for the approval of planned activities, products, processes and equipment?

☐ ☐ ☐

Do you have a procedure for defining responsibility and assigning authority for investigating the EMS for noncompliance and rectifying that noncompliance?

☐ ☐ ☐

Do you have a procedure for identifying, maintaining and controlling environmental records including monitoring and measuring results, training records, audit findings and review reports?

☐ ☐ ☐

Do you have a procedure for controlling all related EMS documents to ensure that: they can be located; they are reviewed, revised and approved regularly; current versions of documents are kept and available where needed; and obsolete documents are removed or otherwise marked accordingly?

☐ ☐ ☐

Do you have a procedure for carrying out periodical audits of the EMS?

☐ ☐ ☐

Chapter 10

Environmental training and awareness

Objective of the chapter

The objective of this chapter is to explain what environmental training and awareness is. This chapter will provide you with the skills necessary to identify what environmental training and awareness is required to maintain a functional EMS. After finishing this chapter, you should be able to answer the following questions:

✓ What is environmental training and awareness?

✓ Why is environmental training and awareness important?

✓ What level of environmental training and awareness is necessary to meet the requirements of ISO 14001 and EMAS?

✓ What written procedures are required for environmental training and awareness?

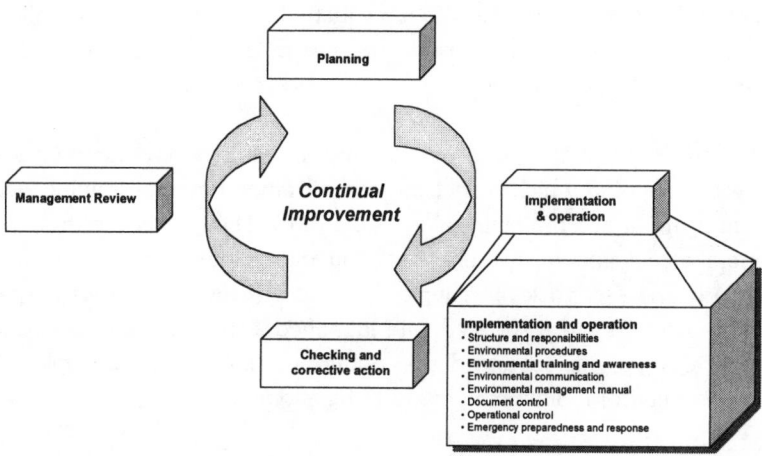

Introduction to training and awareness

For any system to function properly, whether it is a quality management system or an environmental management system, each player within that system must clearly understand their position and how their actions affect the system as a whole.

Consequently, for an EMS to be successful at meeting the policies and objectives set, the various people within that system need to be aware of those policies and objectives and understand how their actions affect the EMS and the overall environmental performance of the site.

If you recall from earlier chapters in the handbook, one of the main purposes of your EMS is to control the environmental aspects of your operation that cause, or could cause, environmental impacts. Similarly, and as discussed in Chapter 2 on the initial environmental review, in order to control your environmental aspects you must first identify what those aspects are. This methodology of 'first identify then address, control or manage' was also used to develop the register of legislation and regulations and the register of aspects and impacts. Logically, as all EMSs require that some training is provided to those involved in the EMS process, the first important step is to assess what training is required. To that end, it is imperative that you are able to clearly identify what training is needed to develop, implement and maintain a functional EMS.

Just as all EMSs are unique to the organizations to which they apply, training needs and requirements will vary from one EMS to another. Essentially, training must be provided to ensure that *all* personnel are aware of the fact that an EMS is being developed and to ensure that *all* personnel are aware of the environmental policy and its importance. Additionally, appropriate training must be provided to all personnel whose work activities are directly associated with any of the significant environmental impacts identified in your initial environmental review and documented in your register of environmental aspects and impacts. Furthermore, appropriate training must be provided to all personnel who have identified roles and responsibilities within the EMS. This training must ensure that the EMS is developed, implemented, audited and maintained properly.

In developing, implementing and maintaining an EMS, many companies provide three levels of training which reflect the overall environmental training needed to develop, implement and maintain a functional EMS. The first level of training is typically general environmental awareness training and an introduction to environmental management. The second level is typically more specific training for all personnel whose work activities are associated with identified significant aspects and impacts. The third level is typically advanced EMS auditor training for those people with identified responsibility for maintaining the EMS developed.

Level 1 training is often referred to as *Environmental Awareness Training*, it should be provided to *all* personnel and would:

- Address the larger global environmental issues such as global warming, the greenhouse effect, acidification and so on and explain why they occur.

- Provide an introduction to environmental management systems and the system being developed at their place of work.

- Develop a clear understanding of the link between the environmental aspects of their work place and identifiable environmental impacts, and subsequent correlation to both global and local environmental issues.

- Address the site's environmental policy, and the importance of conformance to it, and address the objectives and targets set to meet the intentions of the environmental policy.

Level 2 training is often referred to as *Aspects and Impacts Training* and is usually more specific to the aspects and impacts identified in the initial review and documented in the register of environmental aspects and impacts. This training should be provided to *all* personnel whose work activities are associated with or could create a significant environmental impact. This training would:

- Address the environmental impacts of the significant aspects identified in the initial environmental review and documented in the register of environmental aspects and impacts.

- Provide a clear picture of the correlation between aspects and impacts and how individual work activities affect them.

- Develop a clear understanding of the procedures required to control significant aspects identified and why procedures are required.

- Address the importance of adherence to operational procedures and the consequences of procedural noncompliance.

Level 3 training is often referred to as *EMS Training* or *EMS Auditing* and is usually more detailed training on the EMS being developed. It would be provided to *all* personnel who have identified roles and responsibilities in developing, implementing, auditing and maintaining the EMS. This level of training would:

- Address all the system requirements of a functional EMS.

- Address all the roles and responsibilities required to develop, implement and maintain the EMS at your site.

- Address the consequences of failing to comply with the roles and responsibilities defined.

- Address EMS audits and auditing procedures.

In short, training is required to ensure that all people within your EMS are aware of your organization's environmental policy and objectives (intention to improve environmental performance) and the environmental impacts of their actions (control of environmental aspects and minimization of correlating environmental impacts). Training is also required for all personnel whose activities may lead to a significant environmental impact and, lastly, appropriate training is required for all personnel who have an identified role or responsibility for ensuring that the EMS is maintained and meets the requirements of a functional EMS.

Hint

Effective training is not a quick once over but rather a continuing process of human resource development and empowerment.

Certification tip

It is essential that you have a procedure for identifying your training needs and you should be able to substantiate that your training needs have been assessed accordingly. To help assess training needs and provide transparency to the needs assessment, develop a list of personnel who need training, clearly identify what training they need and the programme you used or intend to use to meet those training needs. It cannot be stressed enough that certifying your EMS is not just about developing the right documents and procedures; it requires environmental awareness from all of the employees. Training is an area that is often underestimated and is usually the source of problems during the certification processes.

📄 You must have a written procedure detailing how you identify the training needs required to develop, implement and maintain your EMS.

📄 You must have a written procedure detailing how you train all personnel about the environmental policy and conformance to it, environmental objectives and targets and the EMS at your site.

📄 You must have a written procedure detailing how you train all personnel whose work activities are associated with the significant environmental aspects of your operations as identified in your initial environmental review.

You must have a written procedure detailing how you train all personnel about the consequences of departure from written operating procedures intended to control the environmental aspects of your operations.

You must have a written procedure detailing how you train suppliers, contractors and those acting on your company's behalf to ensure they are aware of, and comply with, your environmental policy.

Your training records should be documented. These documents should be considered controlled documents and be referenced in your environmental management manual.

As with most of the EMS requirements discussed previously, the definitions of and requirements for training differ slightly between ISO 14001 and EMAS. In ISO 14001 this requirement is referred to as *training, awareness and competence* while EMAS uses the term *personnel, communication and training*. In both cases, however, the concept is essentially the same and training on and awareness about the aforementioned EMS concepts and components is a requirement of all the recognized standards.

More precise requirements for training and awareness are discussed in the next section, *Recommendations for successful EMS implementation*.

As awareness and training requirements will vary depending on factors unique to each organization, no generic template has been developed for this section. For examples of what other companies have done, visit the handbook web site at http://www.entropy-international.com/handbook/.

When you think you are finished, you can audit your performance and assess your success by using the check-list in the *Are you ready?* section at the end of this chapter.

Recommendations for successful EMS implementation

✓ You *must* ensure that all personnel receive relevant training for, and are aware of, the development, implementation and maintenance of your EMS.

✓ You *must* ensure that personnel at all levels of operation receive training about, and are aware of, the potential environmental impacts associated with any activities, products or processes with which they are associated.

✓ You *must* ensure that personnel at all levels of operation are aware of the environmental benefits of improved performance with respect to any activities, products or processes with which they are associated.

✓ You *must* ensure that personnel at all levels of operation are aware of your environmental policy and are aware of the importance of compliance with it.

✓ You *must* ensure that personnel at all levels of operation are aware of your objectives and are aware of the importance of compliance with them.

✓ You *must* ensure that personnel at all levels of operation are aware of your environmental procedures and are aware of the importance of compliance with them.

✓ You *must* ensure that all personnel are aware of their roles and responsibilities in meeting the commitments of your environmental policy, objectives and the requirements of your EMS.

✓ You *must* ensure that personnel at all levels of operation are aware of the requirements of your EMS and are aware of the importance of compliance with them.

✓ You *must* ensure that personnel at all levels of operation are aware of the relevant procedures for preparedness and response to any emergency and/or accident situations that they may be associated with.

✓ You *must* ensure that personnel at all levels of operation are aware of the potential consequences of departure from agreed operating procedures.

✓ You *must* maintain a procedure to identify the various and ongoing training needs of your organization with respect to the requirements of your EMS.

✓ You *must* ensure that contractors or others working on behalf of your organization have the appropriate training with respect to the requirements of your EMS.

✓ You *must* ensure that all personnel, contractors and others working on behalf of your organization have the appropriate level of competence, experience and training to minimize the environmental impacts associated with your activities, products and processes and thus comply with the requirements of your EMS.

Are you ready?

<table>
<tr><td colspan="2">

EMS Check-list
Environmental Training and Awareness
page 1 of 2

</td><td>■</td><td>◪</td><td>◨</td></tr>
<tr><td colspan="2"></td><td>Yes</td><td>Partly</td><td>No</td></tr>
</table>

	Yes	Partly	No
Have you ensured that all personnel have received relevant training for, and are aware of, the development, implementation and maintenance of the EMS?	❏	❏	❏
Have you ensured that personnel at all levels of operation have received training about, and are aware of, the potential environmental impacts associated with the activities, products or processes with which they are associated?	❏	❏	❏
Have you ensured that personnel at all levels of operation are aware of the environmental benefits of improved performance with respect to the activities, products or processes with which they are associated?	❏	❏	❏
Have you ensured that personnel at all levels of operation are aware of your organization's environmental policy and are aware of the importance of compliance with it?	❏	❏	❏
Have you ensured that personnel at all levels of operation are aware of the site environmental objectives and are aware of the importance of compliance with them?	❏	❏	❏
Have you ensured that personnel at all levels of operation are aware of the site environmental procedures and are aware of the importance of compliance with them?	❏	❏	❏
Have you ensured that all personnel are aware of their roles and responsibilities in meeting the commitments of your environmental policy, environmental objectives and the requirements of the EMS?	❏	❏	❏
Have you ensured that personnel at all levels of operation are aware of the requirements of the EMS and are aware of the importance of compliance with them?	❏	❏	❏
Have you ensured that personnel at all levels of operation are aware of the relevant procedures for preparedness and response to any emergency and/or accident situations that they may be associated with?	❏	❏	❏

EMS Check-list
Environmental Training and Awareness

page 2 of 2

	■	◩	☐
	Yes	Partly	No

Have you ensured that personnel at all levels of operation are aware of the potential consequences of departure from agreed operating procedures?   ☐ ☐ ☐

Do you maintain a procedure to identify the various and ongoing training needs of your organization with respect to the requirements of your EMS?   ☐ ☐ ☐

Have you ensured that contractors, or others working on behalf of your organization, have the appropriate training with respect to the requirements of the EMS?   ☐ ☐ ☐

Have you ensured that all personnel, contractors and others working on behalf of your organization have the appropriate level of competence, experience and training to minimize the environmental impacts associated with activities, products and processes and thus comply with the requirements of the EMS?   ☐ ☐ ☐

Chapter 11

Environmental communication

Objective of the chapter

The objective of this chapter is to explain what environmental communication is. This chapter will provide you with the skills necessary to identify what environmental communication is required to maintain a functional EMS. After finishing this chapter, you should be able to answer the following questions:

✓ What is environmental communication?

✓ Why is environmental communication important?

✓ What is an environmental statement?

✓ What should be included in an environmental statement?

✓ What environmental communication is necessary to meet the requirements of ISO 14001 and EMAS?

✓ What written procedures are required for environmental communication?

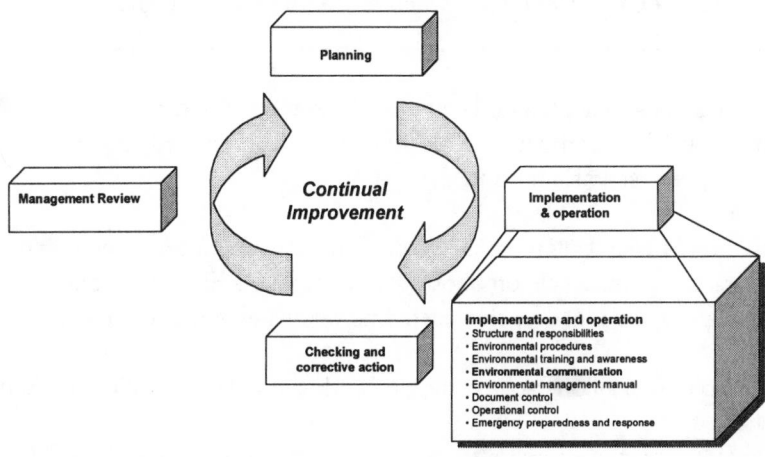

Introduction to environmental communication

As discussed in Chapter 10, it is essential for everyone involved in your EMS to be trained on and aware of your organization's overall environmental intentions and how their individual actions affect the realization of these intentions. Similarly, it is essential to understand that a functional system relies not only upon trained individuals but also communication between them. Furthermore, a functional EMS is not only dependent upon internal communication within the system but also between the system itself and the numerous external stakeholders it affects. These stakeholders would include anyone affected by your environmental impacts or your EMS, such as shareholders, neighbours, clients, suppliers etc.

To this end, environmental communication generally falls into two categories: internal communication and external communication. *Internal communication is the communication between the various levels and functions involved in the development, implementation and maintenance of your EMS.* Internal environmental communication would include things such as: training personnel on the environmental policy; interaction between personnel with identified responsibilities for maintaining the EMS; and informing upper management on changes to or results of your EMS. *External communication is essentially communication with those who are affected by your environmental aspects and/or your EMS.* Your environmental policy is also a form of external communication.

Certification tip

As it is a requirement of ISO 14001 that you make your environmental policy available to the public and communicate your organization's environmental performance, you should be able to indicate clearly how these requirements are met. Additionally, it is good practice to be able to show how you increase the awareness of environmental aspects and impacts, policies, objectives and targets, and environmental management programmes within your company.

You must have a written procedure detailing the method of internal communication about your company's EMS and the environmental aspects and impacts identified as being significant.

You must have a written procedure detailing how you receive, record and respond to external communication about your company's EMS, environmental aspects and impacts identified as significant and overall environmental performance.

Your communication records should be documented and filed for future reference.

As with most of the EMS requirements discussed previously, the definitions of and requirements for communication differ slightly between ISO 14001 and EMAS. However, with the exception of the environmental statement required for EMAS registration, communication is effectively the same thing and is included in each of the EMS requirements for the same reason.

The environmental statement

As mentioned above and in Chapter 1, one of the more fundamental differences between EMAS and ISO 14001 is the EMAS requirement of an environmental statement. As this handbook is meant for those wishing to implement an EMS certifiable to ISO 14001, only a brief summary of the environmental statement is provided here.

In short, the environmental statement should include:

- A description of the activities at your site

- A description of the significant environmental issues at your site

- Figures on pollution emissions, waste generation, raw material consumption and other significant environmental aspects

- Factors regarding environmental performance

- Your environmental policy, and a summary of your environmental management programmes and the management system implemented at the site

- The deadline set for your next statement

- The name of your environmental verifier

> **Note**
>
> *An environmental statement is not the same thing as an environmental policy!*

> **Note**
>
> *An environmental statement is not required for EMAS registration where the verifier considers that the nature and scale of the operations at the site are such that a statement will not be required until completion of a later audit or where few significant changes have occurred since the last statement.*

More precise requirements for communication are discussed in the next section, *Recommendations for successful EMS implementation.*

As communication requirements will vary depending on factors unique to each organization, no generic template has been developed for this section. For examples of what other companies have done, visit the handbook web site at http://www.entropy-international.com/handbook/.

When you think you are finished, you can audit your performance and assess your success by using the check-list in the *Are you ready?* section at the end of this chapter.

Recommendations for successful EMS implementation

✓ Your organization *must* establish and maintain a procedure for receiving, documenting and responding to communications (both internal and external) from relevant interested parties concerning your environmental management system and the environmental aspects and impacts associated with your activities, products and processes.

✓ Your organization's environmental policy *must* be publicly available.

✓ If your organization wishes to be registered to EMAS, you will need to prepare an environmental statement.

✓ The environmental statement *must*:

- Be written after the initial environmental review and after every subsequent audit or audit cycle

- Be clearly written, understandable and intended for the public

- Include a description of your company's activities, products and processes

- Describe the significant environmental issues of relevance to the activities, products and processes

- Provide a summary of the figures on significant emissions, waste generated, raw material consumption and other significant environmental aspects

- Address issues relating to your organization's environmental performance

- Include your environmental policy, a summary of your environmental management programmes and the EMS implemented

- Include deadlines for your next environmental statement

- Include the name of your EMS verifier

✓ Your environmental statement *must* identify the significant changes made since the previous statement.

✓ A simplified environmental statement *must* be prepared annually and shall identify significant changes made since the previous statement.

Are you ready?

<table>
<tr><td colspan="4">

EMS Check-list
Environmental Communication
page 1 of 2

</td><td>■
Yes</td><td>◩
Partly</td><td>☐
No</td></tr>
<tr><td colspan="4">

Have you established and do you maintain a procedure for receiving, documenting and responding to internal and external communication from relevant interested parties concerning your EMS and significant environmental aspects and impacts?

</td><td>☐</td><td>☐</td><td>☐</td></tr>
<tr><td colspan="4">

Is your environmental policy publicly available?

</td><td>☐</td><td>☐</td><td>☐</td></tr>
<tr><td colspan="4">

If you wish to be registered to EMAS, have you prepared an environmental statement?

</td><td>☐</td><td>☐</td><td>☐</td></tr>
<tr><td colspan="7">

If you wish to be registered to EMAS and have prepared an environmental statement, has the environmental statement:

</td></tr>
<tr><td colspan="4">

- Been prepared following your IER or after an audit or audit cycle?

</td><td>☐</td><td>☐</td><td>☐</td></tr>
<tr><td colspan="4">

- Been written in a clear and concise form understandable to the public?

</td><td>☐</td><td>☐</td><td>☐</td></tr>
<tr><td colspan="7">

If you wish to be registered to EMAS and have prepared an environmental statement, does the environmental statement:

</td></tr>
<tr><td colspan="4">

- Include a description of your company's activities, products and processes?

</td><td>☐</td><td>☐</td><td>☐</td></tr>
<tr><td colspan="4">

- Include a description of the significant environmental issues associated with your activities, products and processes?

</td><td>☐</td><td>☐</td><td>☐</td></tr>
<tr><td colspan="4">

- Include a summary of the figures on emissions, waste generated, raw material consumption and other significant environmental aspects?

</td><td>☐</td><td>☐</td><td>☐</td></tr>
<tr><td colspan="4">

- Address other issues important to your environmental performance?

</td><td>☐</td><td>☐</td><td>☐</td></tr>
<tr><td colspan="4">

- Include your environmental policy, a description of your environmental management programmes and your EMS?

</td><td>☐</td><td>☐</td><td>☐</td></tr>
<tr><td colspan="4">

- State a deadline for your next statement and include the name of your EMS verifier?

</td><td>☐</td><td>☐</td><td>☐</td></tr>
</table>

EMS Check-list
Environmental Communication

page 2 of 2

■	◨	◻
Yes	Partly	No

Does your environmental statement address any changes of significance since your previous environmental statement?

☐ ☐ ☐

Is a simplified environmental statement prepared annually and does that statement address significant changes made since your previous statement?

☐ ☐ ☐

Chapter 12

Environmental management manual

Objective of the chapter

The objective of this chapter is to explain what the environmental management manual is. This chapter will provide you with the skills necessary to develop your own environmental management manual. After finishing this chapter, you should be able to answer the following questions:

✓ What is an environmental management manual?

✓ Why is an environmental management manual important?

✓ What must your environmental management manual include to meet the requirements of ISO 14001 and EMAS?

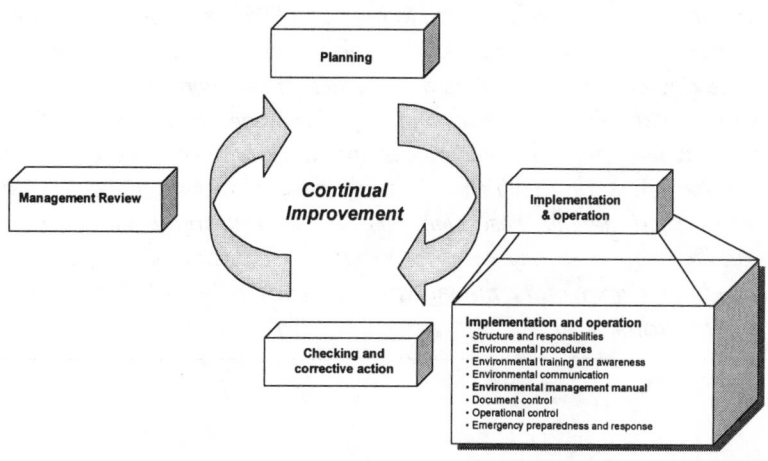

Introduction to the environmental management manual

By now, you should have noticed that the development of a functional EMS is a straightforward, organized and logical process. Similarly, you probably will have noticed that one of the necessities of developing a functional EMS is documentation.

At this point in your EMS implementation process, you should have developed:

- An environmental policy

- A register of aspects and impacts

- A register of legislation and regulations

- Objectives and targets

- Environmental management programmes

- EMS maintenance and operational procedures

For your EMS to be truly functional, this documentation must be straightforward, organized and logical.

Here enters the environmental management manual (EMM). *The environmental management manual is the central tool or reference for the key documents that are required for maintaining and auditing your EMS over time.* The manual can be a single binder containing all your key EMS documents or it can simply be an index referring to the location of those key documents. The choice you make must reflect the needs of your organization and will be dictated largely by existing documents that you will want to incorporate into your emerging EMS.

Hint

The manual does not necessarily need to bind together physically all of your EMS documents but it should provide all the information necessary to identify and locate those documents and explain the interaction between them. For example, the environmental policy, objectives and targets may best be distributed throughout the site while environmental procedures may best be kept closer to the location of the operation to which they apply most. However, make no mistake: your EMM should identify these elements and state clearly where they can be found. Additionally, you may want to consider referencing existing documents and procedures within other management systems to avoid duplicating your management efforts.

Hint

A common referencing system for your controlled documents will help to provide clarity within your EMM and maintain document control. If you do not have an existing methodology try to develop a referencing system that is robust enough to integrate your existing management system documentation (as opposed to being confined to your EMS only). The following table provides a description of the various document references used in the example templates throughout this handbook.

Prefix	Description of the document referenced
AAR001	Environmental management manual amendment and authorization record 001
EP001	Environmental policy 001
AP98001	Audit plan, developed for 1998, plan 001
AR98001	Audit report, developed for 1998, report 001
AS98001	Audit schedule, developed for 1998, schedule 001
CHK001	Checklist 001
EMP001	Environmental management programme 001
NCR001	Nonconformance report 001
OAT001	Objective and targets 001
PROC001	Procedure 001
RAI001	Register of aspects and impacts 001
RLR001	Register of legislation and regulations 001
ESR001	Environmental structure and responsibility 001

The EMM it is not an explicit requirement of either ISO 14001 or EMAS, but for both ISO 14001 and EMAS it is required that you *maintain information* or *documentation* on the core elements of you EMS. Core elements should be seen as those that are essential to meeting the stated intentions expressed in your environmental policy and would include: the environmental policy; environmental objectives and targets; environmental management programmes; procedures, roles and responsibilities within the EMS; and the various interactions between the elements of your EMS. Your EMM should describe how you address these core elements and maintain a functional EMS.

┌───┐
│ ──────── **Certification tip** ────────────────────────────────── │
│   │
│ *A manual containing the key components of your EMS provides transparency* │
│ *during your certification process and clearly identifies areas where environmen-* │
│ *tal documentation has been integrated with other systems such as quality or* │
│ *health and safety. It is helpful to arrange the contents of your manual in line with* │
│ *the sections of ISO 14001 or the contents of this handbook.* │
│   │
└───┘

More precise requirements for an environmental management manual are discussed in the next section, *Recommendations for successful EMS implementation.*

As the physical form of an EMM will vary depending on factors unique to each organization, no generic template has been developed for this section. However, a template has been provided to help you keep track of amendments to the sections of your EMM over time. This template should be placed at the front of each section of your EMM. For examples of what other companies have done, visit the handbook web site at http://www.entropy-international.com/handbook/.

When you think you are finished, you can audit your performance and assess your success by using the check-list in the *Are you ready?* section at the end of this chapter.

Recommendations for successful EMS implementation

✓ You *must* establish and maintain EMS information (in paper or electronic form) that describes the key elements of your EMS and explains how the different components of information interact. This is usually done using an environmental management manual.

✓ Your environmental management manual is one of the most important documents of your EMS and acts as a key instrument for controlling the management system.

✓ Your environmental management manual serves as a central reference point for maintaining and auditing the EMS.

✓ Your environmental management manual should be clear, concise and easy to read and understand.

✓ Your environmental management manual can refer to other forms of information or documentation to avoid duplicating documentation. This is especially true for related documents that may already exist prior to the development of an EMS, such as health and safety or quality management documents.

✓ If your organization already has existing formal management systems, such as a quality management system (QMS), your environmental management manual may best be presented as part of the existing management manual.

✓ Your environmental management manual *should* be prepared and maintained so individual pages can be removed and revised.

✓ Every page in your environmental management manual *should* include:

- The EMS section to which it pertains

- The date it was prepared and the date it was last revised

- When it was authorized and who it was authorized by

- A page number in relation to the total number of pages in that section

✓ Your environmental management manual should include or make reference to:

- A table of contents to indexed sections, starting with a description of how to use the EMM and where copies of it are located and when it is revised

- An introduction including a description of your organization, its activities, products and processes (including a flowchart of the operations)

- An overview of your EMS, its components and how they interact

- Your EMS's organizational structure including organizational charts and diagrams

- A description of environmental responsibilities including job descriptions and authorities

- Your environmental policy

- Your objectives and targets

- Your environmental management programmes (environmental action plans)

- Your operational procedures

- Your emergency plans and documentation

- Your register of significant environmental aspects and impacts

- Your register of legislation and regulations

- The records needed for the implementation and maintenance of your EMS

- The results of EMS audits

- The results of management reviews

Environmental Management Manual

Company Name:	**Document Version:**
Department/Site:	**Issue/Revision Date:**
Updated by:	**Replaces Version:**
Approved by:	**Page of**

EMS amendment and authorization record for:

Version reference	Date of issue	Authorized by	Description of change

				Sample template

Environmental Management Manual

Company Name: United Distillers	**Document Version:**	AAR001
Department/Site: Dailuaine Distillery	**Issue/Revision Date:**	16/07/97
Updated by: Wirral Green	**Replaces Version:**	None
Approved by: Grant Fromage	**Page** 1 **of** 1	

EMS amendment and authorization record for:

Register of environmental aspects and impacts

Version reference	Date of issue	Authorized by	Description of change
RAI001V1	03/05/97	Wirral Green	first issue
RAI001V2	01/04/98	Will Getitdon	new process reviewed and identified significant aspects added to the register
RAI001V3	20/10/99	Wirral Green	register reviewed and updated during EMS audit programme

Are you ready?

	Yes	Partly	No
EMS Check-list **Environmental Management Manual** *page 1 of 2*	■	◪	☐

Have you established and do you maintain EMS information to describe the key components of your EMS and shows how the various pieces of information interact? ☐ ☐ ☐

Is your environmental management manual (EMM) one of the most important documents of your organization's EMS, acting as a key instrument for controlling the EMS? ☐ ☐ ☐

Does your EMM serve as a central reference point for maintaining and auditing the EMS? ☐ ☐ ☐

Is your EMM clear, concise and easy to read and understand? ☐ ☐ ☐

Does your EMM make reference to other forms of information or documentation to avoid duplicating documentation? ☐ ☐ ☐

If you already have existing formal management systems, such as a quality management system (QMS), is your EMM presented as part of the existing management manuals? ☐ ☐ ☐

Is your EMM prepared and maintained so that individual pages can be removed and revised? ☐ ☐ ☐

Does every page in your EMM include:

- The EMS section to which it pertains? ☐ ☐ ☐
- The date it was prepared and the date it was last revised? ☐ ☐ ☐
- When it was authorized and who it was authorized by? ☐ ☐ ☐
- A page number in relation to the total number of pages in that section? ☐ ☐ ☐

EMS Check-list
Environmental Management Manual

page 2 of 2

	■ Yes	◪ Partly	☐ No

Does your EMM include:

- A table of contents to indexed sections, starting with a description of how to use the EMM and where copies of it are located and when it is revised? ☐ ☐ ☐

- An Introduction including a description of your organization, its activities, products and processes – including a flowchart of the operations? ☐ ☐ ☐

- An overview of your organization's EMS, its components and how they interact? ☐ ☐ ☐

- Your EMS's organizational structure including organizational charts and diagrams? ☐ ☐ ☐

- A description of your environmental responsibilities including job descriptions and authorities? ☐ ☐ ☐

- Your environmental policy? ☐ ☐ ☐

- Your objectives and targets? ☐ ☐ ☐

- Your environmental management programmes? ☐ ☐ ☐

- Your EMS and operating procedures? ☐ ☐ ☐

- Your emergency plans and documentation? ☐ ☐ ☐

- Your register of significant environmental aspects and impacts? ☐ ☐ ☐

- Your register of legislation and regulations? ☐ ☐ ☐

- The records needed for the implementation and maintenance of your EMS? ☐ ☐ ☐

- The documentation pertaining to operational control of the activities, products and processes that have, or could have if uncontrolled, a significant impact on the environment? ☐ ☐ ☐

- The results of EMS audits? ☐ ☐ ☐

- The results of management reviews? ☐ ☐ ☐

Chapter 13

Document control

Objective of the chapter

The objective of this chapter is to explain what document control is. This chapter will provide you with the skills necessary to ensure that your level of document control is sufficient to maintain a functional EMS. After finishing this chapter, you should be able to answer the following questions:

✓ What is document control?

✓ Why is document control important?

✓ What is a controlled document?

✓ What is an uncontrolled document?

✓ What level of document control is necessary to meet the requirements of ISO 14001 and EMAS?

✓ What written procedures are required for document control?

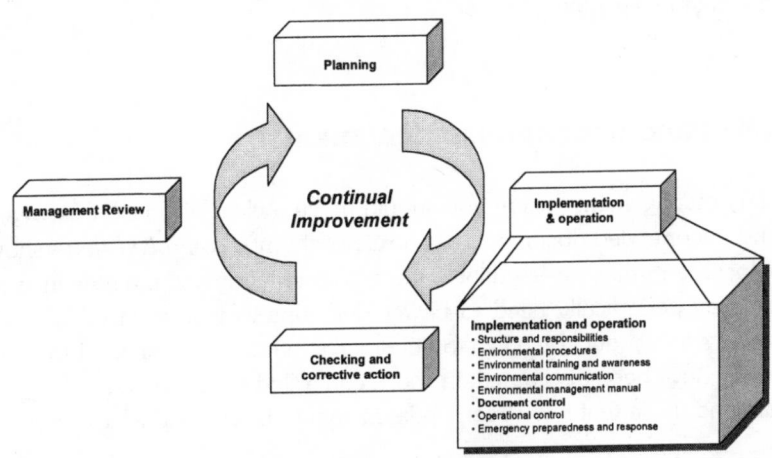

Introduction to document control

By now you are nearing the completion of the implementation stage of your EMS and the elements discussed in the past few chapters should have been seen as fairly logical components of any functional management system.

Document control is no different. In the previous chapter, relating to the environmental management manual, the key point was that a functional management system requires the maintenance of information about that system. Using similar logic, the key point of document control is that if you maintain documented information about a system – whether that system is a quality management system, health and safety system or an EMS – the documents you maintain should be 'controlled'. Essentially, *document control is the set of procedures by which you ensure your EMS documents are organized, current, locatable and 'controlled' in a manner that guarantees their efficacy.*

> **Certification tip**
>
> *It is important to maintain continual control of your environmental documentation by removing or replacing obsolete documents, reviewing, revising, approving and dating new documents issued. Your documentation should include cross references between related documents of your system.*

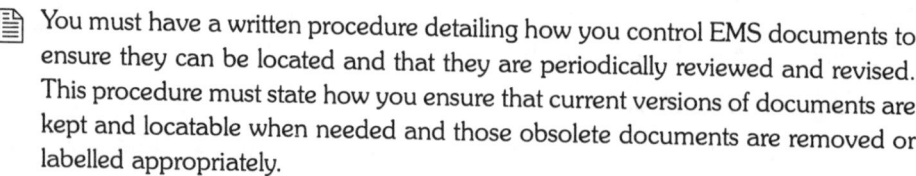 You must have a written procedure detailing how you control EMS documents to ensure they can be located and that they are periodically reviewed and revised. This procedure must state how you ensure that current versions of documents are kept and locatable when needed and those obsolete documents are removed or labelled appropriately.

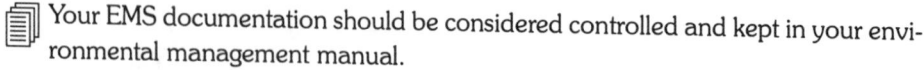 Your EMS documentation should be considered controlled and kept in your environmental management manual.

Controlled and uncontrolled documents

As with any management system you should make a clear distinction between controlled and uncontrolled documents. As a rule of thumb, *controlled status* should be given to documents that are essential to the implementation and maintenance of your EMS and may periodically need to be recalled, updated or replaced. Conversely, *uncontrolled status* should be given to documents that are not intended to be updated once they have been distributed. In short, controlled documents should always be up to date and those that have been updated should be labelled accordingly.

To ensure proper handling of controlled documents it is essential to keep a record of which documents are controlled and where they are located and once a controlled document has been updated, newer versions must replace all known copies of the previous version of that controlled document. You should have a procedure outlining how these documents are referenced, authorized, distributed, stored, retrieved, amended, redistributed and replaced. To avoid confusion, you may find it useful to attach a unique reference number to your controlled documents.

Hint

Needless to say, the more copies of your controlled documents in circulation the more difficult it will be to control them. To ease document control, develop a distribution list and where possible keep the number of controlled documents in circulation to a minimum.

More precise requirements for document control are discussed in the next section, *Recommendations for successful EMS implementation.*

As document control is more of an idea or action and will vary depending on factors unique to each organization, no generic template has been developed for this section. For examples of what other companies have done, visit the handbook web site at http://www.entropy-international.com/handbook/.

When you think you are finished, you can audit your performance and assess your success by using the check-list in the *Are you ready?* section at the end of this chapter.

Recommendations for successful EMS implementation

✓ You *must* maintain a procedure controlling all the documents that are required for the development, implementation and maintenance of your EMS.

✓ You *should* maintain procedures in writing for preparing, storing, issuing, amending and revising documents.

✓ You *must* ensure that all documents that are required for the development, implementation and maintenance of your EMS are legible and clearly marked with the date prepared, date to be revised, document title and document version.

✓ Your document control procedures *must* ensure that all documents that are required for the development, implementation and maintenance of your EMS:

- Are authorized by a top management representative

- Are locatable

- Have a scheduled review period and are revised and authorized according to that review

- Are updated and that updated document versions replace previous and obsolete document versions

- Are updated and provided to all those who require them and in all locations where they are kept

- Are clearly marked as older versions if retained for financial, legal or other reasons

✓ Your document control procedure *should* differentiate between controlled and uncontrolled documents.

Are you ready?

<table>
<tr><td colspan="2">

EMS Check-list
Document control
page 1 of 1

</td><td>■
Yes</td><td>◩
Partly</td><td>□
No</td></tr>
<tr><td colspan="2">Do you maintain procedures controlling all the documents that are required for the development, implementation and maintenance of your EMS?</td><td>❏</td><td>❏</td><td>❏</td></tr>
<tr><td colspan="2">Do you maintain procedures for preparing, storing, issuing, amending and revising environmental management documents?</td><td>❏</td><td>❏</td><td>❏</td></tr>
<tr><td colspan="2">Do you ensure that all documents that are required for the development, implementation and maintenance of your EMS are legible and clearly marked with the date prepared, date to be revised, document title and document version?</td><td>❏</td><td>❏</td><td>❏</td></tr>
<tr><td colspan="2">Do your document control procedures ensure that *all documents* required for the development, implementation and maintenance of your EMS:</td><td></td><td></td><td></td></tr>
<tr><td>●</td><td>Are authorized by a top management representative?</td><td>❏</td><td>❏</td><td>❏</td></tr>
<tr><td>●</td><td>Are locatable?</td><td>❏</td><td>❏</td><td>❏</td></tr>
<tr><td>●</td><td>Have a scheduled review period and are revised and authorized according to that review?</td><td>❏</td><td>❏</td><td>❏</td></tr>
<tr><td>●</td><td>Are updated and that updated document versions replace previous and obsolete document versions?</td><td>❏</td><td>❏</td><td>❏</td></tr>
<tr><td>●</td><td>Are updated and provided to all those who require them and in all locations where they are kept?</td><td>❏</td><td>❏</td><td>❏</td></tr>
<tr><td>●</td><td>Are clearly marked as older versions if they are retained for financial, legal or other reasons?</td><td>❏</td><td>❏</td><td>❏</td></tr>
<tr><td colspan="2">Do you differentiate between controlled and uncontrolled documents?</td><td>❏</td><td>❏</td><td>❏</td></tr>
</table>

Chapter 14

Operational control

Objective of the chapter

The objective of this chapter is to explain what operational control is. This chapter will provide you with the skills necessary to ensure that your level of operational control is sufficient to maintain a functional EMS. After finishing this chapter, you should be able to answer the following questions:

✓ What is operational control?

✓ Why is operational control important?

✓ What level of operational control is necessary to meet the requirements of ISO 14001 and EMAS?

✓ What written procedures are required for operational control?

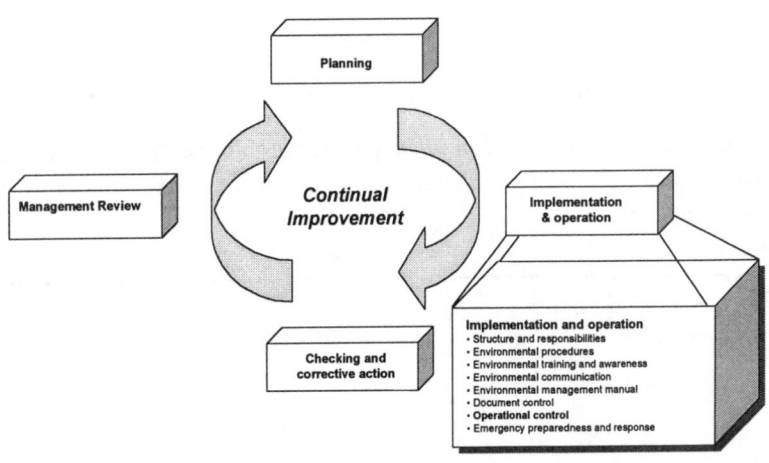

Introduction to operational control

While almost all issues that surround the systematic approach to environmental management have been at the centre of one debate or another, two assumptions remain largely uncontested. The first is that all companies, sites and organizations have an impact on the environment. The second is that by systematically controlling the activities, products and processes to minimize their correlating environmental impacts an organization will improve its overall environmental performance.

Essentially then, improved environmental performance should be seen as contingent upon improved control of activities, products and processes (operations) that cause an identified environmental impact and, in doing so, minimize the significant environmental impacts identified.

Therefore, just as document control was the set of procedures by which you ensure your EMS documents are 'controlled', *operational control is the set of procedures that ensure your operations (aspects) are controlled*. Operational control consequently improves environmental performance by controlling aspects and minimizing correlating environmental impacts caused by those operations.

Hint

Operational control is established through the development and observation of procedures detailing how an operation is to be completed. Operational control is an essential management tool that will ensure your environmental policy and objectives and targets can be met. Operational control should cover all operations that are associated with any of the environmental impacts deemed significant in your initial environmental review process.

Certification tip

It is important to be able to show that you have operational procedures for all activities or processes that are directly associated with an environmental impact deemed significant.

You must have a written procedure detailing how you identify significant environmental aspects and correlating impacts associated with your operations (activities, products and processes).

 You must have a written procedure detailing operational criteria for the activities, products and processes that are associated with environmental aspects identified as being significant.

More precise requirements for operational control are discussed in the next section, *Recommendations for successful EMS implementation*.

As operational control is more of an idea or action and will vary depending on factors unique to each organization, no generic template has been developed for this section. For examples of what other companies have done, visit the handbook web site at http://www.entropy-international.com/handbook/.

When you think you are finished, you can audit your performance and assess your success by using the check-list in the *Are you ready?* section at the end of this chapter.

Recommendations for successful EMS implementation

✓ You *must* identify and attempt to control physically all the activities, products and processes that have been associated with the significant environmental impacts identified in your initial environmental review.

✓ Operational control *must* include operating procedures defining the manner in which your organization will conduct the activities, functions and processes (aspects) which have, or could have if uncontrolled, identified significant environmental impacts.

✓ You *must* prepare operational procedures for all activities, products and processes where the absence of such instructions would, or could, lead to a significant environmental impact and thus compromise your environmental policy.

✓ You *must* prepare operational procedures for all personnel working for or on behalf of your organization if their activities, products and processes lead to, or could lead to, a significant environmental impact.

✓ You *must* monitor with the aim to control all significant environmental impacts of your activities, products and processes.

✓ Your operational control *must* include the approval (based on environmental criteria) of all planned activities, products, processes and procurements.

✓ Your operational control *must* include performance criteria for all activities, products and processes that have, or could have if uncontrolled, a significant impact on the environment.

Are you ready?

EMS Check-list **Operational Control** *page 1 of 1*	Yes	Partly	No
Do you identify and attempt to control physically all the activities, products and processes that have been associated with the significant environmental impacts identified in the initial environmental review?	❏	❏	❏
Do your control measures include documented operating procedures defining the manner in which employees will conduct the activities and processes which have, or could have if uncontrolled, significant environmental impacts?	❏	❏	❏
Have you prepared operational procedures for all activities, products and processes where the absence of such procedures would, or could, lead to a significant environmental impact?	❏	❏	❏
Have you prepared operational procedures for all personnel working for or on behalf of your organization if their activities, products or processes lead to, or could lead to, a significant environmental impact?	❏	❏	❏
Do you monitor, with the aim to control, all significant environmental impacts of your organization's activities, products and processes?	❏	❏	❏
Do your operational controls include the approval of all planned activities, products, processes and procurements?	❏	❏	❏
Do your operational control measures include performance criteria for all activities, products and processes that have, or could have if uncontrolled, a significant impact on the environment?	❏	❏	❏

Chapter 15

Emergency preparedness and response

Objective of the chapter

The objective of this chapter is to explain what emergency preparedness is. This chapter will provide you with the skills necessary to identify what level of emergency preparedness is required to maintain a functional EMS and develop methodologies for preventing accidents and emergencies and controlling them when and if they occur. After finishing this chapter, you should be able to answer the following questions:

✓ What is emergency preparedness?

✓ Why is emergency preparedness important?

✓ What level of emergency preparedness is necessary to meet the requirements of ISO 14001 and EMAS?

✓ What written procedures are required for emergency preparedness?

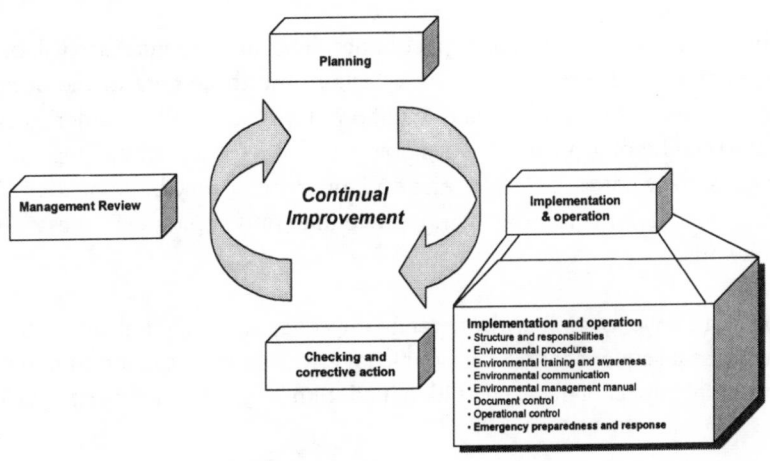

Introduction to emergency preparedness and response

By now your EMS is nearly complete. You have:

✓ Identified and documented your significant environmental aspects and impacts

✓ Established the intention to improve environmental performance

✓ Developed a systematic approach to minimize those impacts

✓ Provided a structure for that system

✓ Identified the responsibilities required to maintain that structure

✓ Developed a procedural approach to ensure consistency of the structured system and its requisite documentation

✓ Established the importance of awareness about the system and its overall intentions and provided training where such awareness lacks or compromises the integrity of the intentions of the system

✓ Implemented procedural control of all the operations that have, or could have, a significant direct or indirect impact on the environment

In short, you have developed a procedural and systematic methodology to control the activities, products and processes that cause an identified significant environmental impact. Furthermore, you have supported that control with an infrastructure to maintain and ensure its efficacy.

If we lived in a perfect world, EMS implementation would now be complete. However, we do not and the final component that must be added to complete the implementation process is *a procedural and systematic methodology for preventing accidents and emergencies and controlling them when and if they occur.*

As the nature and environmental impact of accidents and emergencies will of course vary depending on the organization in question and the nature of the accident or emergency, there is no standardized definition of an 'accident' or 'emergency situation' that would be of value when developing an EMS. However, the important issue is not necessarily what 'is' or 'is not' an accident or emergency but that your EMS has procedural mechanisms for identifying, preventing and coping with these situations when they arise.

As discerned above, overall environmental improvement is contingent on the identification, and subsequent control, of significant environmental aspects to minimize the environmental impact of the aspects identified. Similarly, a functional EMS will ensure that accidents and emergency situations, or the potential for them, are identified, prevented where possible and controlled in the event that they occur.

Hint

When developing procedures for emergency situations, such as fire risks or potential spillages, it can be helpful to use a map of your site to identify the potential areas of concern and the steps that should be taken to minimize such risks. You should assess the potential for accidents and emergencies that may be associated with the more salient aspects of your operations. These would include the significant aspects already identified in the initial review, areas where accidents and emergencies have happened in the past, areas where toxic, hazardous or special chemicals are used, and areas where worker health and safety may be of concern.

Certification tip

Your certifier will be looking for evidence that you have appropriate procedures in place for dealing with possible environmental incidents and potential emergency situations. At a minimum you should consider accidental discharges to land or water, emissions to atmosphere and related environmental and ecosystem impacts.

 You must have a written procedure detailing the process you use for identifying the likelihood of accidents and emergencies. This procedure should include details about the prevention techniques for identified risks and the minimization, control and mitigation of such accidents or emergencies if they occur. The efficacy of this procedure should be tested regularly.

More precise requirements for accident and emergency situations are discussed in the next section, *Recommendations for successful EMS implementation.*

As accident and emergency situations will vary depending on factors unique to each organization, no generic template has been developed for this section. For examples of what other companies have done, visit the handbook web site at http://www.entropy-international.com/handbook/.

When you think you are finished, you can audit your performance and assess your success by using the check-list in the *Are you ready?* section at the end of this chapter.

Recommendations for successful EMS implementation

✓ You *must* establish and maintain procedures to identify the potential for accidents and emergency situations associated with your activities, products and processes.

✓ You *must* establish and maintain procedures to respond appropriately to, and minimize the environmental impact of, accident and emergency situations.

✓ You *must* regularly review, and revise when necessary, your accident and emergency procedures.

✓ You *must* regularly test the efficacy of your accident and emergency response procedures and revise them when necessary.

Are you ready?

	Yes	Partly	No
EMS Check-list **Emergency Preparedness and Response** *page 1 of 1*	■	◪	▨
Have you established and do you maintain procedures to identify the potential for accidents and emergency situations associated with your activities, products and processes?	❑	❑	❑
Have you established and do you maintain procedures to respond appropriately to, and minimize the environmental impact of, accident and emergency situations?	❑	❑	❑
Do you regularly review and revise, when necessary, your accident and emergency procedures?	❑	❑	❑
Do you regularly test the efficacy of your accident and emergency response procedures and revise them when necessary?	❑	❑	❑

Chapter 16

Monitoring and measuring

Objective of the chapter

The objective of this chapter is to explain what monitoring and measuring is. This chapter will provide you with the skills necessary to identify what monitoring and measuring is required to maintain a functional EMS. After finishing this chapter, you should be able to answer the following questions:

✓ What is monitoring and measuring?

✓ Why is monitoring and measuring important?

✓ What level of monitoring and measuring is necessary to meet the requirements of ISO 14001 and EMAS?

✓ What written procedures are required for monitoring and measuring?

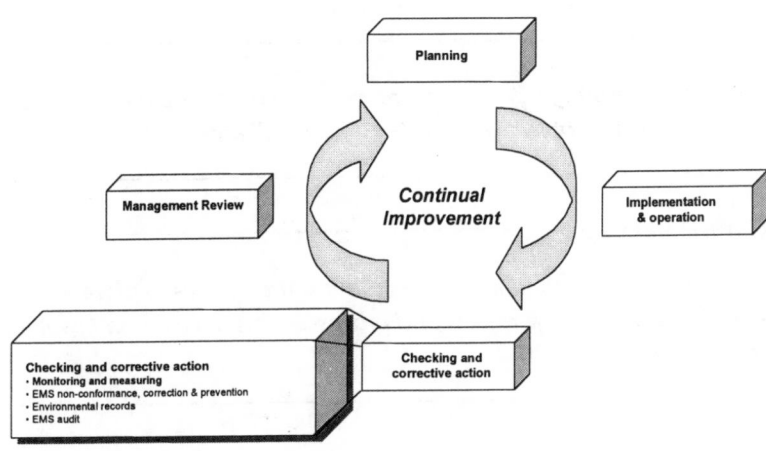

Introduction to monitoring and measuring

As you can see from the diagram on the previous page, you should now have completed the 'Planning' and 'Implementation' stages of the 'Stages of ISO 14001 implementation' diagram (the 'Plan' and 'Do' stages of the 'Cycle of improvement' diagram in Chapter 1). At this point you may think you are done. However, as this is not just 'a system' but rather a system based on continual improvement, your EMS must have an internal mechanism to track and guarantee improvement. Thinking back to Chapter 2 you will recall that the first step in improving environmental performance was to identify your environmental impacts and develop a clear understanding of what you are doing and where you will improve. You are now beginning the 'Checking and corrective action' stage of your EMS implementation (the 'Check' stage of the 'Cycle of improvement' diagram in Chapter 1).

An EMS must provide a comparative reference or benchmark over time by which it assesses progress toward its stated goal of environmental improvement. *Monitoring and measuring is the means by which an organization identifies its progress towards the minimization of the environmental impact of its activities, products and processes.*

In the United Distillers case study, for example, the stated goal was overall environmental improvement. This was to be met, in part, by 'reducing water use by 50% by the year 2000'. Consequently, in order to track the performance of this objective and target they would have to both monitor and measure the water used in that period of time.

In short, if an environmental impact is to be minimized by subsequent control of the environmental aspects from which it stems then all the activities, functions, products or processes (aspects) that are associated with that impact should be regularly monitored and measured.

Hint

Monitoring and measuring will generate records. Records should be used to track progress towards stated objectives and targets (see Chapter 17).

Certification tip

You will need to show that your organization has procedures in place to monitor and measure the activities, products and processes that have a significant impact on the environment.

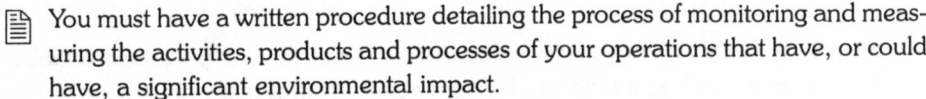 You must have a written procedure detailing the process of monitoring and measuring the activities, products and processes of your operations that have, or could have, a significant environmental impact.

Monitoring and measuring records should be documented. These documents should be considered controlled documents and be referenced in your environmental management manual.

More precise requirements for monitoring and measuring are discussed in the next section, *Recommendations for successful EMS implementation.*

As monitoring and measuring will vary depending on the environmental impacts of each organization, no generic template has been developed for this section. For examples of what other companies have done, visit the handbook web site at http://www.entropy-international.com/handbook/.

When you think you are finished, you can audit your performance and assess your success by using the check-list in the *Are you ready?* section at the end of this chapter.

Recommendations for successful EMS implementation

✓ In order to improve corporate environmental performance, you *must* establish and maintain procedures to monitor the activities, products and processes that are associated with a significant environmental impact (as indicated in your initial environmental review). This enables your organization to control those activities, products and processes and reduce the associated environmental impact.

✓ Similarly, in order to improve corporate environmental performance, you *must* establish and maintain procedures to measure the impact of these aforementioned activities, products and processes.

✓ You *must* document what is to be monitored and measured.

✓ You *must* record the results of monitoring and measuring. This will provide a benchmark for improved environmental performance and enable you to evaluate environmental performance with respect to your environmental policy and objectives and targets.

✓ You *must* establish and maintain procedures for the calibration (and recording the result of such calibration) of all monitoring and measuring equipment.

✓ You *must* establish and document acceptance criteria for the results of monitoring and measuring activities.

✓ You *must* establish and document the action to be taken if the results of monitoring and measuring are unsatisfactory.

✓ You *must* evaluate and document the validity of previous records if it is determined that monitoring and measuring systems have been malfunctioning.

Are you ready?

	Yes	Partly	No
EMS Check-list **Environmental Monitoring and Measuring** *page 1 of 1*	■	◩	☐
Have you established and do you maintain procedures to monitor the activities, products and processes that are associated with a significant environmental impact?	☐	☐	☐
Have you established and do you maintain procedures to measure the impact that results from the aforementioned activities, products and processes?	☐	☐	☐
Have you documented what is to be monitored and measured?	☐	☐	☐
Do you record the results of monitoring and measuring to provide a benchmark for improved environmental performance?	☐	☐	☐
Have you established and do you maintain procedures for the calibration (and recording the result of such calibration) of all monitoring and measuring equipment?	☐	☐	☐
Have you established and documented acceptance criteria for the results of monitoring and measuring activities?	☐	☐	☐
Have you established and documented the action to be taken if the results of monitoring and measuring are unsatisfactory?	☐	☐	☐
Have you evaluated and documented the validity of previous records if it has been determined that monitoring and measuring systems have been malfunctioning?	☐	☐	☐

Chapter 17

EMS nonconformance, correction and prevention

Objective of the chapter

The objective of this chapter is to explain what nonconformance, correction and prevention are. This chapter will provide you with the skills necessary to identify nonconformances and apply the corrective and preventative action required to maintain a functional EMS. After finishing this chapter, you should be able to answer the following questions:

✓ What are nonconformance, correction and prevention?

✓ Why are nonconformance, correction and prevention important?

✓ What level of correction and prevention is necessary to meet the requirements of ISO 14001 and EMAS?

✓ What written procedures are required for nonconformance, correction and prevention?

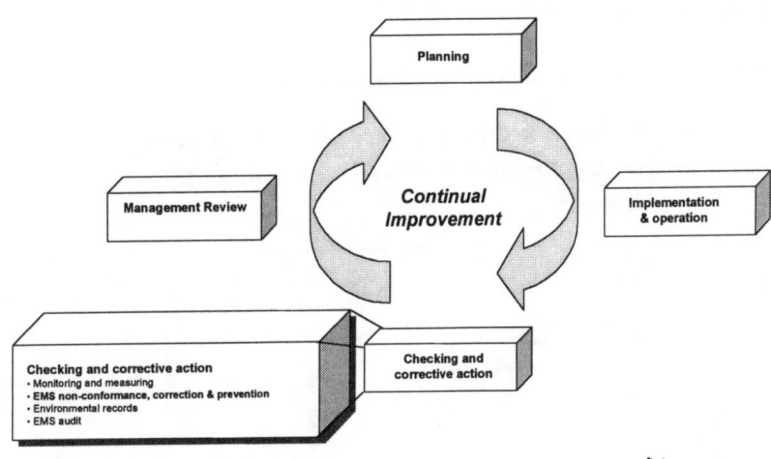

Introduction to nonconformance, correction and prevention

If you recall from Chapter 14, operational control is established by following procedures that ensure your operations are planned and conducted in a manner that minimizes their environmental impact. Minimizing impact consequently improves environmental performance. Therefore, if you have successfully developed procedural control of all the activities, products and processes that cause a significant environmental impact, your system should meet the intentions of your policy, your objectives and targets, and guarantee improved environmental performance.

However, it is unlikely that this is either the case or possible on first attempt at developing a functional EMS. This is because, in most newly developed EMSs, the operational control required to minimize all your significant environmental impacts is often underestimated or insufficient. Fear not, however, as this next component of your EMS is intended to help you identify areas of nonconformance, correct them and prevent them from reoccurring.

Nonconformance is the situation where essential components of your EMS are absent or dysfunctional or where there is insufficient control of your activities, products or processes to the extent that these absences compromise your policy, objectives and targets, management programmes and the functionality of your EMS.

Correction is the act of developing or improving where nonconformance has been identified. Just as objectives and targets have environmental management programmes (action plans) to ensure that you meet the objectives and targets set, corrective action should similarly have documented action plans that must be completed to ensure corrective action is taken and that it is effective.

Prevention is the act of ensuring that nonconformance will not reoccur. Just as you would for any form of prevention, the prevention of nonconformance requires an understanding of what caused the nonconformance and ensuring that cause of the nonconformance is avoided in the future.

Hint

Action plans for correction and prevention should be developed and implemented according to, and commensurate with, the magnitude of nonconformance identified.

Certification tip

You will need to show to your certifier that your EMS audits are capable of identifying nonconformances within your EMS and that corrective and preventative action can be implemented and followed up to ensure its effectiveness.

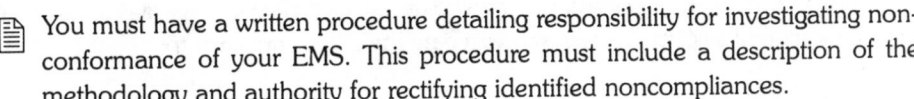 You must have a written procedure detailing responsibility for investigating nonconformance of your EMS. This procedure must include a description of the methodology and authority for rectifying identified noncompliances.

More precise requirements for nonconformance, correction and prevention are discussed in the next section, *Recommendations for successful EMS implementation.*

To develop your own action plans for correction and prevention you can use the blank and sample templates later in this chapter or on the floppy disk at the back of this book. For examples of what other companies have done, visit the handbook web site at http://www.entropy-international.com/handbook/.

When you think you are finished, you can audit your performance and assess your success by using the check-list in the *Are you ready?* section at the end of this chapter.

Recommendations for successful EMS implementation

✓   Your EMS and all of its components, documents and procedures *must* conform to your environmental policy, objectives and targets, and the requirements of the management standard from which you have designed your management system.

✓   In the event of nonconformance with your environmental policy, objectives and targets, and the requirements of the standard to which you have designed your management system, you *must* investigate the noncompliance to:

- Determine what caused the noncompliance.

- Develop a plan for correction for the noncompliance.

- Determine what preventative measures should be taken.

- Ensure preventative measures are effective.

- Ensure that any procedures affected by the corrective action taken, are revised accordingly.

✓   You *must* establish and maintain procedures that define the person(s) responsible for, and the authority they have, for investigating, correcting, mitigating and preventing nonconformances.

✓   You *must* ensure that any corrective, mitigating, or preventative actions taken in light of nonconformances are appropriate to the nature and scale of the associated environmental impact or potential for environmental impact of that nonconformance.

✓   You *should* document and maintain a record of noncompliances.

Nonconformance Report Form

Blank template

Company Name:	**Document Version:**
Department/Site:	**Issue/Revision Date:**
Manager Responsible:	**Replaces Version:**
Approved by:	**Page of**

Nonconformance related to:

Areas affected:	**Audit plan reference:**
	Auditor:

Nonconformance description

Corrective action to be taken

Agreed completion date for corrective action to be taken:

Signed:

             (manager)   **(auditor)**

Follow-up action

Corrective action   **Signed:**

completed on: **(auditor)**

Nonconformance Report Form	*Sample template*

Company Name: United Distillers	**Document Version:** NCR001V1
Department/Site: Dailuaine Distillery	**Issue/Revision Date:** 05/01/98
Manager Responsible: Wirral Green	**Replaces Version:** None
Approved by: Grant Fromage	**Page** 1 **of** 1

Nonconformance related to: Awareness of environmental policy

Areas affected: Cask filling staff	**Audit plan reference:** AP9801
Dark grains plant staff	**Auditor:** Will B. Frank

Nonconformance description

Five of the employees were not aware of Dailuaine's environmental policy and subsequently, were not aware of the environmental policy's relevance to their place of work. This is a direct nonconformance with ISO 14001 requirements for environmental policy.

Corrective action to be taken

- Group discussion/training to be engaged regarding the environmental policy and its role at the site to be provided to all employees in cask filling and the dark grains plant.
- The policy's relevance to specific job descriptions should be made clear to all employees in each area.

Agreed completion date for corrective action to be taken: 01/03/98

Signed: *Wirral Green* *Will B. Frank*
   (manager)   **(auditor)**

Follow-up action

Reassessment of staff members in each area using check-list CHK01.

Corrective action completed on: 20/02/98   **Signed:** *Will B. Frank*
                                          (auditor)

Are you ready?

	Yes	Partly	No
EMS Check-list **Nonconformance, Correction and Prevention** *page 1 of 1*	■	◩	◻

Does your EMS and all its components, documents and procedures conform with the environmental policy, objectives and targets and the requirements of the management standard from which your management system has been designed? ☐ ☐ ☐

In the event of nonconformance, have you investigated the nonconformance to:

- Determine what caused the nonconformance? ☐ ☐ ☐

- Determine what correction is required for the nonconformance? ☐ ☐ ☐

- Determine what preventative measures should be taken? ☐ ☐ ☐

- Ensure that preventative measures implemented are effective? ☐ ☐ ☐

- Ensure procedures affected by the corrective taken are revised accordingly? ☐ ☐ ☐

Have you established and do you maintain procedures that define the person(s) responsible for, and the authority they have, for investigating, correcting, mitigating and preventing nonconformances? ☐ ☐ ☐

Do you ensure that any corrective, mitigating or preventative actions are appropriate to the nature and scale of the associated environmental impact or potential for environmental impact of that nonconformance? ☐ ☐ ☐

Do you document and maintain a record of nonconformances? ☐ ☐ ☐

Chapter 18

Environmental records

Objective of the chapter

The objective of this chapter is to explain what environmental records are. This chapter will provide you with the skills necessary to identify what environmental records are required to maintain a functional EMS. After finishing this chapter, you should be able to answer the following questions:

✓ What are environmental records?

✓ Why are environmental records important?

✓ What environmental records are necessary to meet the requirements of ISO 14001 and EMAS?

✓ What written procedures are required for environmental records?

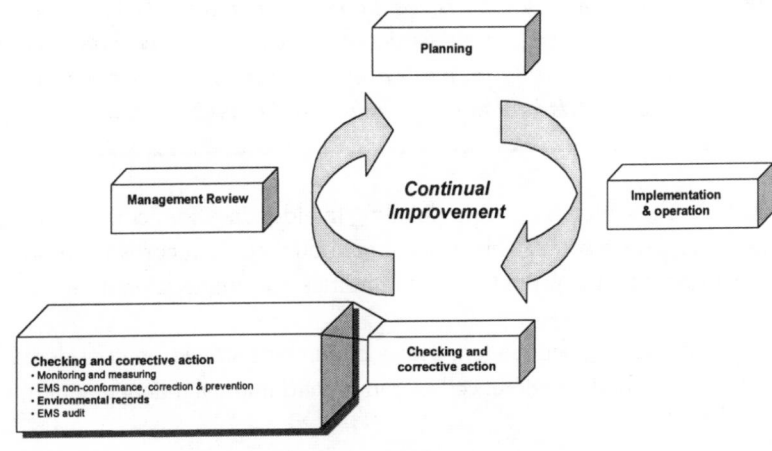

Introduction to environmental records

Another requirement of a functional EMS is maintenance of what are referred to as EMS records. At first glance it may appear there is confusion and redundancy with the need for manuals, document controls and now records. However, remembering that you are now in the 'check' stage of the EMS process may provide some clarity.

The 'documents' discussed previously in the handbook are those that are essential components of a functional EMS. *These documents, such as the policy, the objectives and targets, the environmental management programmes and the procedures are those that are fundamental to the EMS itself. EMS records refer to these EMS documents but also include the documents that contain data by which you benchmark the performance of your EMS.* EMS records include things like water-use measurements, waste generation records, and air emission monitoring results and are the result of monitoring and measuring. Records should measure the operations (aspects) associated with significant environmental impacts identified in your initial review and should be used to evaluate progress in meeting your objectives and targets.

Hint

If an activity, product or process (aspect) has been identified as being the cause of a significant environmental impact, that aspect should consequently be monitored (measured) regularly to serve as an indicator of environmental performance.

Certification tip

Records provide evidence that your EMS is being maintained. You will want to keep records for environmental aspects and their correlating impacts, product composition information, permits, legislative and regulatory requirements, monitoring and measurement, inspections, calibration and maintenance, incidents, complaints, nonconformance, corrective action, supplier/contractor data as well as information related to your EMS audits and management reviews. It is wise to maintain a list of all records kept and regularly update that list when necessary.

You must have a written procedure detailing the identification, control and maintenance of all EMS records. This procedure should cover records for monitoring and measuring, training provided, audit findings and management reviews.

All records should be documented. These documents should be considered controlled documents and be referenced in your environmental management manual.

More precise requirements for EMS records are discussed in the next section, *Recommendations for successful EMS implementation.*

As the EMS records kept will vary depending on each organization's EMS, no generic template has been developed for this section. For examples of what other companies have done, visit the handbook web site at http://www.entropy-international.com/handbook/.

When you think you are finished, you can audit your performance and assess your success by using the check-list in the *Are you ready?* section at the end of this chapter.

Recommendations for successful EMS implementation

✓ You *must* document and record various indicators of performance to track and evaluate environmental improvement in relation to your environmental policy, objectives and targets, and the requirements of your EMS.

✓ You *must* develop and retain all environmental records required for the successful development, implementation and maintenance of your EMS.

✓ While records will differ from company to company depending on what you intend to measure performance against, it is *recommended* that your environmental records include:

- Information on all of your activities, products and processes (aspects) associated with identified significant environmental impacts

- Information on all relevant environmental legislation, regulations or other requirements that are applicable to your organization (register of environmental legislation and regulations)

- Information on the environmental significance of your activities, products and processes

- Information on environmental training provided in your organization

- Information on internal and external complaints related to your EMS or overall environmental performance

- Information on EMS audits and reviews

- Information on suppliers, contractors and others acting on behalf of your organization who may affect the operation of your EMS and your overall environmental performance

- Information on your emergency plans, procedures and response activities

- Information on the inspection and maintenance of your monitoring and measuring equipment

✓ You *must* establish and maintain procedures for the identification, maintenance, retention and disposal of the environmental records you keep.

✓ Your environmental records *must* be legible, identifiable, dated and clearly linked to the activity, product or process (aspect) to which they are associated.

✓ Your environmental records *must* be easy to locate.

Are you ready?

EMS Check-list **Environmental Records** *page 1 of 1*	■ Yes	◨ Partly	☐ No
Do you track and evaluate environmental improvement using recorded indicators of environmental performance?	☐	☐	☐
Have you developed and retained all environmental records required for the successful development, implementation and maintenance of your EMS?	☐	☐	☐
Do your environmental records include:			
• Information on all of your activities, products and processes (aspects) associated with identified significant environmental impacts?	☐	☐	☐
• Information on all of the environmental legislation, regulations or other requirements that are applicable to your organization?	☐	☐	☐
• Information on the environmental significance of your activities, products and processes?	☐	☐	☐
• Information on environmental training provided in your organization?	☐	☐	☐
• Information on internal and external complaints related to your EMS or overall environmental performance?	☐	☐	☐
• Information on EMS audits and reviews?	☐	☐	☐
• Information on suppliers, contractors and others acting on your behalf who may affect your EMS and your overall environmental performance?	☐	☐	☐
• Information on your emergency plans, procedures and response?	☐	☐	☐
• Information on the inspection and maintenance of your monitoring and measuring equipment?	☐	☐	☐
Have you established, and do you maintain, procedures to identify, maintain, retain and dispose of environmental records kept?	☐	☐	☐
Are your records locatable, legible, identifiable, dated and clearly linked to the activity, product or process to which they are associated?	☐	☐	☐

Chapter 19

EMS audit

Objective of the chapter

The objective of this chapter is to explain what EMS audits are. This chapter will provide you with the skills necessary to perform an EMS audit and identify what level of auditing is required to maintain a functional EMS. After finishing this chapter, you should be able to answer the following questions:

✓ What are EMS audits?

✓ Why are EMS audits important?

✓ What level of EMS auditing is necessary to meet the requirements of ISO 14001 and EMAS?

✓ How should you conduct an EMS audit?

✓ How should you report your EMS audit findings?

✓ What written procedures are required for EMS audits?

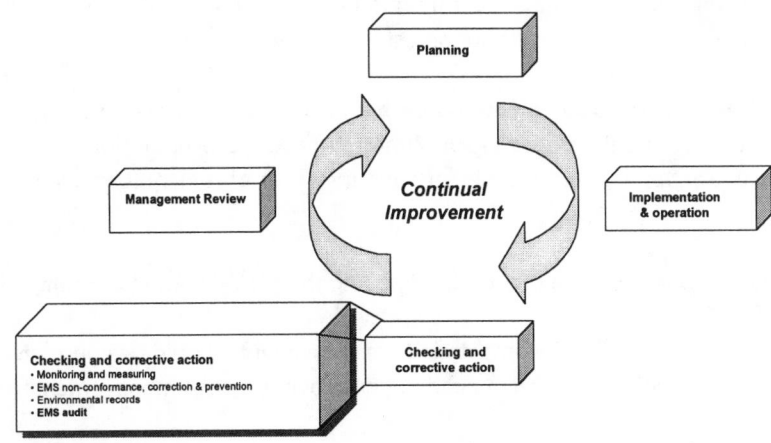

Introduction to the EMS audit

Having now established mechanisms for monitoring and measuring, controlling non-conformance, and environmental records, the checking and corrective action process culminates with the EMS audit. At this point you have implemented the fundamental pieces of your EMS and must now assess its efficacy and improve where efficacy is identified as lacking.

An EMS audit, like a QMS audit or financial audit, is the process by which you assess whether your system meets a set of previously defined criteria. In this case, the criteria are the requirements of the ISO 14001 standard. The audit assesses how well your EMS meets those criteria (i.e. are you doing what you say you are doing?)

When performing an EMS audit, it is imperative that you plan, perform and report your audit in a manner that will ensure efficacy of not only the audit itself but ultimately the efficacy of the recommendations for EMS improvement that the audit generates. In doing so, the following are of critical importance when planning, performing and reporting the results of your EMS audit:

1   **The audit plan**. Be sure to have thorough audit plans that include a schedule, scope and responsibility for the audit. Where possible relate the scope and frequency of your audits to past performance and your significant environmental aspects and impacts.

2   **Internal auditing procedures**. Be sure that your audit plans cover the procedures and methodology for conducting your audits. Your audit procedures should address the competence, experience, training, independence of auditors and the checks and verifications to be performed.

3   **Audit report**. Be sure you address your methodology for reporting and disseminating the findings and recommendations of your audit to appropriate personnel. An identified distribution list for the report is often useful.

4   **Audit follow-up**. Be sure you establish an action programme for implementing audit recommendations and ensure that you have a procedure for ensuring that corrective action is taken for all identified instances of nonconformance.

You must have a written procedure detailing the periodic audit of your EMS.

Audit results should be documented. These documents should be considered controlled documents and be referenced in your environmental management manual.

Planning your EMS audit

The first step in developing an EMS audit is the preparation of your audit plan. Essentially, your audit plan describes how you will conduct your audit (i.e. what you will audit, who will audit, when you will audit and how you will audit). In planning your EMS audit you should define the scope, objectives and resources necessary for conducting the audit.

- The audit scope determines the extent and boundaries of the audit (i.e. what locations or parts of the EMS are to be audited, which activities are to be audited, how the audit will be reported). The activities associated with significant environmental aspects should be audited more frequently.

- In most cases the objective of your EMS audit will be to determine whether the EMS conforms to your planned arrangements and the requirements of ISO 14001. This overall objective can be further broken down within your audit plans to address specific requirements: i.e. to audit the environmental policy, objectives and targets, environmental management programmes, and structure and responsibility.

- Selecting an appropriate team is critical for performing a successful EMS audit. As a rule of thumb you should try to assemble a team that has formal training in auditing and questioning techniques and brings together a diverse range of skills and technical knowledge relevant to the EMS being audited. It is also important to ensure your audit team exhibits a high degree of independence and objectivity in meeting the requirements of the audit plan.

> **Note**
>
> *There are many ways to plan your EMS audit. You could audit each activity, product or process one by one to assess conformance to your EMS standard requirements. Conversely, you could audit specific components of the standard against your EMS to assess whether they are present and managing your activities, products and processes as planned. The most important thing is that you design an audit methodology that works for you and ensures that the entire system is audited over time. Not all components of your EMS must be audited simultaneously or with the same frequency.*

> **Hint**
>
> *To put your plan into practice it is necessary to develop a schedule and document your audit plan accordingly. See the templates later in this chapter for an example of an audit schedule that covers the entire EMS and documented audit plans that investigate the environmental policy.*

Conducting your EMS audit

The typical audit is conducted under the responsibility of a lead auditor, in consultation with the audit team. The lead auditor assigns auditing tasks to various audit team members, such as specific EMS elements or activities, and instructs them (according to the defined audit plan) on the appropriate procedures to be followed.

During the audit it is the lead auditor's responsibility to organize and compile the documents that are produced by the audit team. These documents include:

- Forms for documenting and supporting audit evidence and audit findings

- Procedures and checklists used for evaluating EMS elements

- Meeting and interview records

A typical audit usually consists of the following five steps:

1   **Opening meeting.** Your audit should begin with an opening meeting. The purpose of this meeting is to discuss the audit plan and provide introductions to the people involved. This generally involves:

- Discussing the scope, objectives, audit plan and timetable

- Explaining the assessment methodologies that will be used in the audit

- Appointing relevant contact persons

- Ensuring that the necessary resources are available

- Promoting the active staff participation of the company or site being audited

- Briefing on relevant site, safety and emergency procedures for the audit team

2   **Collecting evidence.** It is the task of the auditor to collect sufficient evidence to be able to assess whether or not the EMS conforms to the audit criteria. This evidence is gathered by conducting interviews, examining documents and observation of on-site activities. It is not common to conduct tests or measurements (i.e. auditors do not usually generate new information). Remember, it is an audit of the system not of environmental performance!

An essential tool for collecting evidence is the check-list. It provides a list of written questions designed to assess conformance with a given set of audit criteria. Check-lists are particularly useful in interviews or while observing on-site activities. They provide a record of the audit and an auditor can use them irrespective of familiarity with the company. See the templates later in this chapter for examples of check-lists you can use in your audit.

Other tools that can be used to generate audit evidence include audit trails, fishbone diagrams and sampling.

- Audit trails generally involve tracing the paperwork related to a particular component of your EMS. This ensures that all of the relevant documents, records and procedures related to that component are present and locatable, that they are functioning as they should and that the 'trail of documents' is complete and unbroken. For example, you could follow the use of a toxic chemical at your site from its point of purchase through to its final disposal to ensure all documents and procedures involved are up to date and being followed correctly.

- A fishbone diagram is an effective way of identifying probable root causes of a particular unwanted effect or nonconformance. By using this diagram, all of the potential causes are exhaustively listed in order to determine the most probable cause of the unwanted effect.

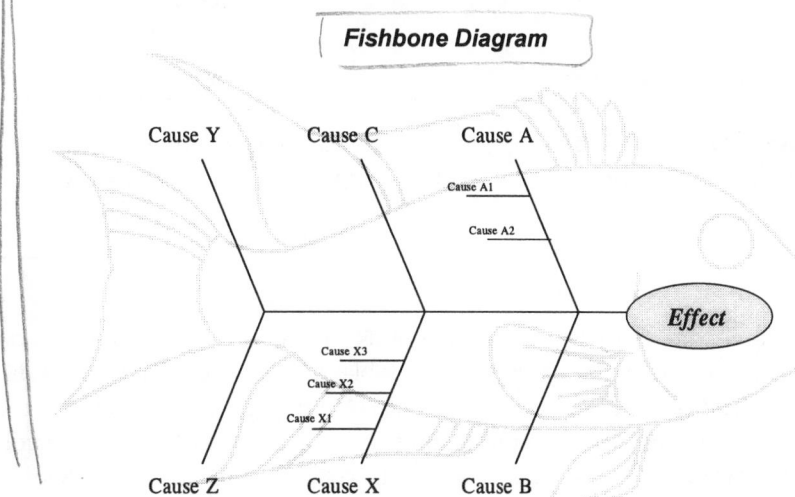

- Although rarely used, in some instances it may be necessary to take your own samples for independent testing and results analysis.

3 **Audit findings**. Having conducted interviews, observed site activities and collected relevant information, to identify areas of nonconformance and generate findings, you will need to evaluate this evidence using the criteria identified in your audit plan.

At this point you will undoubtedly encounter inconsistencies between standard EMS requirements and the present state of affairs at the site being audited. These inconsistencies could be trivial, such as poor language in a procedure, to serious, such as the lack of crucial operating procedures. The key here is to differentiate between minor and inconsequential discrepancies and major system nonconformances.

As you will recall from Chapter 17, nonconformance is the situation where essential components of your EMS are absent or dysfunctional or where there is insufficient control of your significant environmental aspects that then compromises your policy, objectives and targets, and management programmes. For example, failure to follow an operational procedure for controlling a significant environmental aspect or the absence of a controlled EMS document should be treated as a major nonconformance.

While you should document all observed discrepancies as findings in your audit report, serious nonconformances should be dealt with in a special manner. A template for documenting and correcting nonconformances has been included later in this chapter. All nonconformances should be documented in a clear and concise manner and supported by sufficient audit evidence to ensure reliability.

> **Hint**
>
> *It is useful to record findings of conformity to illustrate the number of checks that have been conducted and provide some relativity to any nonconformances found.*

---

4 **Closing meeting**. Before compiling the audit report, the findings of the audit should be presented to the personnel responsible for the component being audited (the auditee). The purpose of this meeting is to provide an opportunity for the auditee to put forward further information (which may be accepted by the auditor as additional evidence) and change the status of the current findings if necessary. In short, the closing meeting provides a forum for challenging findings, resolving disagreements and developing plans for necessary corrective action.

> **Hint**
>
> *As you will always be dealing with a limited amount of available information, you should wherever possible discuss your findings with the auditee to confirm their validity.*

---

5 **Report preparation**. The lead auditor is responsible for co-ordinating the preparation of the report and for its accuracy and completeness. The issues to be covered in the report are normally described in the audit plan.

Your audit report should be objective and care must be taken to avoid conflicts of interest between the auditors preparing the report and those directly responsible for the component being audited. A typical audit report should include:

- Details of the audit team

- The scope, objectives and assessment criteria of the audit

- The audit plan followed

- A summary of the audit process

- Time-frame of the audit

- Any confidentiality arrangements

- Audit recommendations, which are clearly based on the audit findings

- Signature of the lead auditor

- A distribution list for the report

More precise requirements for an EMS audit are discussed in the next section, *Recommendations for successful EMS implementation*. Additionally, you may wish to consult the ISO member body closest to you (contact information provided in Annex 4) for an official copy of the following guidelines:

Guidelines for environmental auditing – General principles (ISO 14010:1996)

Guidelines for environmental auditing – Audit procedures – Auditing of environmental management systems (ISO 14011:1996)

As there are many ways to conduct an EMS audit, it is difficult to provide a full and complete audit template suitable for all companies. However, the blank and sample templates later in this chapter (or on the floppy disk at the back of this book) will help you through your audit and make your EMS auditing activities more transparent for your certifiers. For examples of what other companies have done, visit the handbook web site at http://www.entropy-international.com/handbook/.

When you think you are finished, you can audit your performance and assess your success by using the check-list in the *Are you ready?* section at the end of this chapter.

Recommendations for successful EMS implementation

✓ You *must* have clearly defined audit programme(s) for conducting regular EMS audits.

✓ You *must* have clearly defined audit procedures for conducting EMS audits.

✓ Your audit programme(s) and procedures *must* determine whether your EMS conforms to the requirements of the relevant standard or regulation to which you subscribe.

✓ Your audit programme(s) and procedures *must* determine whether your EMS conforms to your stated environmental policy, objectives and targets, management programmes and regulatory requirements.

✓ Your audit programme(s) and procedures *must* ascertain whether the EMS has been implemented correctly and is sufficiently maintained.

✓ Your audit programme(s) and procedures *must* include the reporting of your audit results to the top management representatives.

✓ Your audit programme(s) and resulting audit schedule *must* be based on the environmental significance of your activities and your results from previous audits or reviews.

✓ Your audit procedures *must* include a clearly defined audit scope specifying the interval covered by the audit and the activities, areas and EMS standards to be taken into account.

✓ Your audit procedures *must* refer to the frequency that EMS audits are to be conducted.

✓ Your audit procedures *must* include the responsibilities and conditions associated with managing, performing and reporting the results of your audits.

✓ Your EMS audit *must* include an assessment of relevant data that can be used to evaluate performance objectively.

✓ Your audit programme(s) and procedures *should* ensure auditor's competence.

✓ Your EMS audit *must* be conducted by people with appropriate training and experience in:

- The areas and activities being audited

- Environmental management systems

- Technical and legislative requirements

- The particular skills needed to achieve the EMS audit objectives

✓ The resources and time allotted for your EMS audit *must* be proportional to the scope and objectives you have set.

✓ Top management *must* support your EMS audit.

✓ Your auditors *must* be adequately independent of the activities and areas being audited to ensure an unbiased assessment.

✓ Your EMS audit(s) *must* be planned and prepared to ensure that the roles and responsibilities of auditors, management and staff are clearly understood and that sufficient resources are provided to conduct the audit.

✓ Preparation for an EMS audit *must* include:

- An understanding of your organization's activities, products and processes

- Familiarity with the EMS established in your organization

- An analysis of the results from previous audits or reviews

✓ To evaluate environmental performance, EMS audit activities *must* include:

- Discussions or interviews with the staff on site

- Investigation of the site operations and related equipment

- A review of relevant EMS documentation such as procedures or records

✓ The following phases *should* be included in your EMS audit process:

- An understanding of the EMS

- A review of the strengths and weaknesses

- Gathering of audit evidence

- Formation of audit findings

- Formation of audit recommendations

- Reporting of audit findings and recommendations

✓ Your EMS audit *should* be conducted in accordance with accepted auditing principles such as those outlined in ISO 14010, ISO 14011 or Annex II of EMAS.

✓ Your EMS audit *must* include the preparation of a written report by the auditors to ensure the provision of findings and recommendations at the end of each audit.

✓ The findings and results of the your EMS audit *must* be made known to top management.

✓ The fundamental objectives of your EMS audit report *are*:

- To document the extent of the audit

- To provide management with a 'gap analysis' of the organization's conformance to the relevant EMS requirements

- To provide management with a review of conformance to the environmental policy and a report on environmental performance in light of planned arrangements

- To provide management with a report on the efficacy of the procedures for monitoring and controlling environmental aspects and impacts at the site

- To identify the need for improvements and corrective action

✓ You *should* have an appropriate procedure in place to ensure that the recommendations and areas of identified corrective action in the audit are followed up.

EMS Audit Schedule

Company Name:

Department/Site:

Updated by:

Approved by:

Document Version:

Issue/Revision Date:

Replaces Version:

Page   of

Reference	Audit to cover	Jan Feb Mar	Apr May Jun	Jul Aug Sep	Oct Nov Dec

	Blank template
Internal EMS Audit Plan	

Company Name:	**Document Version:**
Department/Site:	**Issue/Revision Date:**
Updated by:	**Replaces Version:**
Approved by:	**Page of**

Activity being audited:

Additional Auditors:	**Audit Frequency:**
	Audit Schedule Ref:
	Audit Plan Ref:

Components of the EMS audit plan

Signed:

                    (Managing Director)   **(Lead Auditor)**

EMS Audit Check-list

Blank template

Company Name:	**Document Version:**
Department/Site:	**Issue/Revision Date:**
Updated by:	**Replaces Version:**
Approved by:	**Page of**

Activity being audited:

Methodology:	**Audit Plan Ref:**
	Auditor:

Questions & criteria	Yes	Part	No	Comments

EMS Audit Nonconformance Report Form	*Blank template*

Company Name: **Document Version:**

Department/Site: **Issue/Revision Date:**

Updated by: **Replaces Version:**

Approved by: **Page**   **of**

Nonconformance related to:

Areas affected: **Audit plan reference:**

     Auditor:

Nonconformance description

Corrective action to be taken

Agreed completion date for corrective action to be taken:

Signed:

     (manager)   **(auditor)**

Follow-up action

Corrective action   **Signed:**

completed on: **(auditor)**

Internal EMS Audit Report

Company Name:	**Document Version:**
Department/Site:	**Issue/Revision Date:**
Updated by:	**Replaces Version:**
Approved by:	**Page of**

Activity being audited:

Additional Auditors:	**Audit Schedule Ref:**
	Audit Plan Ref:

Findings of the audit

Recommendations

Signed:

       (Managing Director)   **(Lead Auditor)**

EMS Audit Schedule				Sample template

Company Name: United Distillers **Document Version:** AS98001V1
Department/Site: Dailuaine Distillery **Issue/Revision Date:** 15/12/97
Updated by: Wirral Green **Replaces Version:** None
Approved by: Grant Fromage **Page** 1 **of** 1

Reference	Audit to cover	Jan Feb Mar	Apr May Jun	Jul Aug Sep	Oct Nov Dec
AP9801	Environmental policy	X		X	
AP9802	Environmental aspects	X		X	
AP9803	Legislation and regulations	X		X	
AP9804	Objectives and targets	X		X	
AP9805	Environmental management programmes	X		X	
AP9806	Structure and responsibility	X			
AP9807	Training, awareness and competence		X		
AP9808	Communication		X		
AP9809	EMS manual and documentation		X		
AP9810	Document control		X		
AP9811	Operational control			X	
AP9812	Emergency preparedness and response	X		X	
AP9813	Monitoring and measurement			X	
AP9814	Nonconformance, correction and prevention	X		X	
AP9815	Records				X
AP9816	EMS audits				X
AP9817	Management review				X

Internal EMS Audit Plan	*Sample template*

Company Name: United Distillers	**Document Version:** AP98001V1
Department/Site: Dailuaine Distillery	**Issue/Revision Date:** 15/12/97
Updated by: Wirral Green	**Replaces Version:** None
Approved by: Grant Fromage	**Page** 1 **of** 1

Activity being audited: Environmental policy

Additional Auditors:	**Audit Frequency:** 12 months
Will B. Frank, Jim Nasium	**Audit Schedule Ref:** AS98001V1
	Audit Plan Ref: AP98001V1

Components of the EMS audit plan

1. Use checklist CHK001 to interview 10 of the 27 employees around the site regarding the environmental policy and their knowledge of it. Ensure that an adequate mix of job descriptions is represented.

2. Use checklist CHK002 to evaluate whether the Dailuaine is practising what it has committed to in the environmental policy.

3. Use checklist CHK003 to evaluate the environmental policy against the requirements of ISO 14001 and EMAS.

4. Identify all findings and recommendations in the final report for the audit referenced AP98001.

Enclose completed check-lists and nonconformance forms in the annex of the report.

Signed: *Grant Fromage* *Wirral Green*

(**Managing Director**) (**Lead Auditor**)

EMS Audit Check-list	*Sample template*

Company Name: United Distillers   **Document Version:** CHK001V1
Department/Site: Dailuaine Distillery   **Issue/Revision Date:** 05/01/98
Updated by: Wirral Green   **Replaces Version:** None
Approved by: Grant Fromage   **Page** 1 **of** 1

Activity being audited: Staff knowledge of environmental policy

Methodology: Questions to be asked of 10 employees on the site to assess their knowledge of the environmental policy.

Audit Plan Ref: AP9801V1
Auditor: Will B. Frank

Questions & criteria	Yes	Part	No	Comments
1. Are you aware of the company's environmental policy?	5		5	Nonconformance to be raised
2. What is the purpose of the environmental policy for Dailuaine?	5		5	Did not know
3. How does the environmental policy apply to the work that you do?	3	1	6	Did not know
4. Do you think the policy is followed on site?	9		1	
5. Do you think the current policy is good enough for the site?	8		2	

EMS Audit Check-list	Sample template

Company Name: United Distillers	**Document Version:** CHK002V1
Department/Site: Dailuaine Distillery	**Issue/Revision Date:** 05/01/98
Updated by: Wirral Green	**Replaces Version:** None
Approved by: Grant Fromage	**Page** 1 **of** 2

Activity being audited: Environmental policy and correlating links to improved environmental performance

Methodology: Questions to be asked in order to assess fulfilment of commitments made in the environmental policy.	**Audit Plan Ref:** AUD001V1
	Auditor: Will B. Frank

Questions & criteria	Yes	Part	No	Comments
1. Has Dailuaine shown commitment to the protection and enhancement of the environment by fulfilling and implementing the environmental policy?	X			
2. Has Dailuaine shown commitment to the protection and enhancement of the environment by meeting legal requirements?	X			
3. Has Dailuaine shown commitment to the protection and enhancement of the environment by considering environmental issues in all operations at the site?	X			
4. Are environmental responsibilities held by the general manager and the environmental manager?	X			
5. Have environmental issues and the views of interested parties, employees and the local community been taken into account in strategic decisions affecting the environment?		X		could develop a finding
6. Has Dailuaine aimed to minimize environmental impacts through use of best available technology and practice (in particular effluent to water, emissions to air and generation of waste)?	X			
7. Is Dailuaine monitoring and reducing water and energy use?		X		develop a finding regarding use of water

EMS Audit Check-list				*Sample template*

Company Name: United Distillers	**Document Version:** CHK002V1
Department/Site: Dailuaine Distillery	**Issue/Revision Date:** 05/01/98
Updated by: Wirral Green	**Replaces Version:** None
Approved by: Grant Fromage	**Page** 2 **of** 2

Activity being audited: Environmental policy and correlating links to improved environmental performance

Methodology: Questions to be asked in order to assess fulfilment of commitments made in the environmental policy.	**Audit Plan Ref:** AUD001V1
	Auditor: Will B. Frank

Questions & criteria	Yes	Part	No	Comments
8. Has Dailuaine ensured that objectives, targets and action plans are determined for reducing and controlling all significant environmental impacts caused by the operations?		X		could develop a finding
9. Are objectives and targets regularly revised with the aim to improve continually the environmental performance of the site?		X		develop a finding regarding setting of objectives and targets
10. Are all new projects evaluated for their environmental impact before sign off?	X			
11. Is a dialogue maintained to ensure that all employees are aware of, and participate in, the environmental work at the site?		X		could develop a finding regarding awareness and participation
12. Is the environmental management system reviewed at regular intervals?	X			
13. Is Dailuaine managing its environmental issues according to the ISO 14001 EMS standard?	X			
14. Does Dailuaine distillery strive for continual improvement and prevention of pollution by actively engaging in environmental projects to control and reduce environmental impacts?	X			
15. Is a proactive and open attitude to environmental issues maintained?		X		could develop a finding

EMS Audit Check-list

Sample template

Company Name: United Distillers   **Document Version:** CHK003V1
Department/Site: Dailuaine Distillery   **Issue/Revision Date:** 05/01/98
Updated by: Wirral Green   **Replaces Version:** None
Approved by: Grant Fromage   **Page** 1 **of** 2

Activity being audited: Environmental policy and meeting of EMS requirements

Methodology: Questions to be asked in order to assess the environment policy's conformance with the requirements in ISO 14001.

Audit Plan Ref: AP98001V1
Auditor: Jim Nasium

Questions & criteria	Yes	Part	No	Comments
1. Does the environmental policy state your organization's principles and intentions in relation to its overall environmental performance?	X			
2. Is the environmental policy appropriate to the nature, scale and environmental impacts of your organization's activities, functions, products and processes?	X			
3. Does the environmental policy include a commitment to continual improvement and the prevention of pollution based on an acceptable methodology?	X			
4. Does the environmental policy include a commitment to comply with relevant environmental legislation, regulations and other requirements to which your organization subscribes?	X			
5. Is the environmental policy documented, implemented, maintained, reviewed and made known to all employees?		X		Nonconformance issued for awareness
6. Does top management endorse the environmental policy?	X			Policy signed by Grant Fromage
7. Is the environmental policy available to the public?		X		

EMS Audit Check-list				*Sample template*

Company Name: United Distillers	**Document Version:**	CHK003V1
Department/Site: Dailuaine Distillery	**Issue/Revision Date:**	05/01/98
Updated by: Wirral Green	**Replaces Version:**	None
Approved by: Grant Fromage	**Page 2 of 2**	

Activity being audited: Environmental policy and meeting of EMS requirements

Methodology: Questions to be asked in order to assess the environment policy's conformance with the requirements in ISO 14001.	**Audit Plan Ref:** AP98001V1
	Auditor: Jim Nasium

Questions & criteria	Yes	Part	No	Comments
8. Does the policy provide the framework for setting and reviewing environmental objectives and targets?	X			
9. Is the environmental policy clear, concise and written in non-technical language interpretable by both internal and external parties?	X			
10. Does the environmental policy include a commitment to the development and implementation of an environmental management system?	X			
11. Does the environmental policy include a commitment to the development of, and adherence to, corporate standards in the absence of legislation?			X	Nonconformance for which finding and recommendation should be made
12. Does the environmental policy embody a life cycle approach to the environmental impacts of your organization's activities, functions, products and processes?			X	Nonconformance for which finding and recommendation should be made

EMS Audit Nonconformance Report Form

Sample template

Company Name:	United Distillers	**Document Version:**	NCR001V1
Department/Site:	Dailuaine Distillery	**Issue/Revision Date:**	05/01/98
Updated by:	Wirral Green	**Replaces Version:**	None
Approved by:	Grant Fromage	**Page** 1 **of** 1	

Nonconformance related to: Awareness of environmental policy

Areas affected:	Cask filling staff	**Audit plan reference:** AP98001V1
	Dark grains plant staff	**Auditor:** Will B. Frank

Nonconformance description

Five of the employees were not aware of Dailuaine's environmental policy and, subsequently, were not aware of the environmental policy's relevance to their place of work. This is a direct nonconformance with ISO14001 requirements for environmental policy.

Corrective action to be taken

- Group discussion/training to be engaged regarding the environmental policy and its role at the site to be provided to all employees in cask filling and the dark grains plant.

- The policy's relevance to specific job descriptions should be made clear to all employees in each area.

Agreed completion date for corrective action to be taken: 01/03/98

Signed: *Wirral Green* *Will B. Frank*

(manager) **(auditor)**

Follow-up action

Reassessment of staff members in each area using check-list CHK01.

Corrective action completed on: 20/02/98

Signed: *Will B. Frank*

(auditor)

Internal EMS Audit Report

Company Name: United Distillers	**Document Version:** AP98001V1
Department/Site: Dailuaine Distillery	**Issue/Revision Date:** 23/01/98
Updated by: Wirral Green	**Replaces Version:** None
Approved by: Grant Fromage	**Page** 1 **of** 2

Activity being audited: Environmental policy

Additional Auditors:	**Audit Schedule Ref:** AS98001V1
Will B. Frank, Jim Nasium	**Audit Plan Ref:** AP98001V1

Findings of the audit

1. The environmental policy has not been made known to all employees and represents a nonconformance with the requirements of ISO 14001. Nonconformance form NCR001 has been raised.

2. The site environmental policy is currently not readily available to the public but could be if asked for. This is an observation that does not completely conform to the specific requirements of ISO 14001.

3. The environmental policy does not include a commitment to the development of, and adherence to, corporate standards in the absence of legislation. This is purely an observation related to best practice.

4. The environmental policy does not embody a life cycle approach to the environmental impacts of the organization's activities, products and processes. This is purely an observation related to best practice.

5. Environmental issues and the views of interested parties, employees and the local community are only partially taken into account in strategic decisions affecting the environment. This is a minor nonconformance with the stated policy.

6. The use of water is not being reduced at present. This is a minor nonconformance with the stated policy conditions (specifying the measurement and reduction of water and energy).

7. The site has not ensured that objectives, targets and action plans are determined for reducing and controlling all significant environmental impacts caused by the operations although the most significant are being controlled at present.

8. Current objectives and targets have a legislation-based focus and do not specifically aim to improve continually the environmental performance of the site. This is an observation that does not completely conform to the stated policy.

9. While very open, the current attitudes toward environmental issues could not be described as proactive. This is an observation that does not completely conform to the stated policy.

Internal EMS Audit Report	Sample template

Company Name: United Distillers	**Document Version:** AP98001V1
Department/Site: Dailuaine Distillery	**Issue/Revision Date:** 23/01/98
Updated by: Wirral Green	**Replaces Version:** None
Approved by: Grant Fromage	**Page** 2 **of** 2

Activity being audited: Environmental policy

Additional Auditors:	**Audit Schedule Ref:** AS98001V1
Will B. Frank, Jim Nasium	**Audit Plan Ref:** AP98001V1

Recommendations

1. The nonconformance form NCR001 regarding staff awareness of the environmental policy should be followed up as agreed and documented with the lead auditor.

2. The site environmental policy should be made readily available to the public and distributed to local residents within a five-mile radius of the site's operations.

3. The inclusion of clauses in the environmental policy specifying adherence to corporate standards in the absence of legislation and the embodiment of a life cycle approach to environmental impacts should be considered.

4. Greater effort must be made to take into account the views of interested parties, employees and the local community when making strategic decisions affecting the environment.

5. A water reduction programme should be put in place to meet the commitment made in the environmental policy.

6. Objectives, targets and action plans should be developed for reducing and controlling all of the significant environmental impacts caused by the operations as outlined in the initial environmental review.

7. In setting objectives and targets, Dailuaine should move from a legislation-based focus to a focus on continual improvement of environmental performance incorporating proactive initiatives.

8. The environmental policy does not embody a life cycle approach to the environmental impacts of the organization's activities, products and processes. This is purely an observation related to best practice.

Signed: *Grant Fromage*   *Wirral Green*
        (Managing Director)   **(Lead Auditor)**

EMS Check-list
EMS Audit

	Yes	Partly	No

Are you

Do you have clearly defined audit programme(s) for carrying out regular EMS audits? ❑ ❑ ❑

Do you have clearly defined audit procedures for conducting EMS audits? ❑ ❑ ❑

Do your audit programme(s) and procedures determine whether your EMS conforms to the requirements of the relevant EMS standard or regulation to which you subscribe? ❑ ❑ ❑

Do your audit programme(s) and procedures determine whether your EMS conforms to your stated environmental policy, objectives and targets, management programmes and regulatory requirements? ❑ ❑ ❑

Do your audit programme(s) and procedures determine whether the EMS has been correctly implemented and sufficiently maintained? ❑ ❑ ❑

Do your audit programme(s) and procedures cover the reporting of your audit results to the top management representatives? ❑ ❑ ❑

Are your audit programme(s) and resulting audit schedule based on the environmental significance of your activities and your results from previous audits or reviews? ❑ ❑ ❑

Do your audit procedures include a clearly defined audit scope specifying the interval covered by the audit and the activities, areas and EMS standards to be taken into account? ❑ ❑ ❑

Do your audit procedures refer to the frequency of which EMS audits are to be conducted? ❑ ❑ ❑

ready?

Do your audit procedures include the responsibilities and conditions associated with managing, performing and reporting the results of your audits? ❑ ❑ ❑

EMS Check-list
EMS Audit

page 2 of 4

	Yes	Partly	No

Do your EMS audits include an assessment of relevant data that can be used to objectively evaluate performance? ❏ ❏ ❏

Do your audit programme(s) and procedures ensure auditor's competence? ❏ ❏ ❏

Are your EMS audits conducted by people with appropriate training and experience in:

- The areas being audited? ❏ ❏ ❏

- Environmental management practices? ❏ ❏ ❏

- Technical and legislative requirements? ❏ ❏ ❏

- The particular skills needed to achieve the EMS audit objectives? ❏ ❏ ❏

Are the resources and time allocated for your EMS audit proportional to the scope and objectives you have set? ❏ ❏ ❏

Does top management support your EMS audit? ❏ ❏ ❏

Are your auditors adequately independent of the activities they audit to ensure an unbiased assessment? ❏ ❏ ❏

Are your EMS audits planned and prepared to ensure that the roles and responsibilities of auditors, management and staff are clearly understood and that sufficient resources are provided to conduct the audit? ❏ ❏ ❏

Does the preparation for your EMS audits include:

- An understanding of your organization's activities, products and processes? ❏ ❏ ❏

- Familiarity with the EMS established in your organization? ❏ ❏ ❏

- An analysis of the results from previous audits or reviews? ❏ ❏ ❏

EMS Check-list
EMS Audit

page 3 of 4

| | ■ Yes | ◩ Partly | ☐ No |

Do your on-site EMS audit activities included:

- Discussions or interviews with staff on site? ☐ ☐ ☐

- Investigation of the site operations and related equipment? ☐ ☐ ☐

- A review of relevant EMS documentation such as procedures or records? ☐ ☐ ☐

Does your EMS audit process include:

- Gaining an understanding of the EMS? ☐ ☐ ☐

- A review of the strengths and weaknesses? ☐ ☐ ☐

- Gathering audit evidence? ☐ ☐ ☐

- Formation of audit findings? ☐ ☐ ☐

- Formation of audit conclusions? ☐ ☐ ☐

- Reporting audit findings and recommendations? ☐ ☐ ☐

Have (are) your EMS audits been conducted in accordance with accepted auditing principles such as those outlined in ISO 14010, ISO 14011 or Annex II of EMAS? ☐ ☐ ☐

Do your EMS audits include the preparation of a written report by the auditors to ensure the provision of findings and recommendations at the end of each audit? ☐ ☐ ☐

Are the findings and recommendations of your EMS audits made known to top management? ☐ ☐ ☐

EMS Check-list
EMS Audit

■ ◧ ☐
Yes Partly No

page 4 of 4

Are the fundamental objectives of your organization's written EMS audit report:

- To document the extent of the audit?   ☐ ☐ ☐

- To provide management with 'gap analysis' of the organization's conformance to the relevant EMS requirements ?   ☐ ☐ ☐

- To provide management with a review of conformance to the environmental policy and a report on environmental performance in light of planned arrangements?   ☐ ☐ ☐

- To provide management with a report on the efficacy of the procedures for monitoring and controlling environmental aspects and impacts at the site?   ☐ ☐ ☐

- To demonstrate the need for improvements and corrective action?   ☐ ☐ ☐

Do you have an appropriate procedure in place to ensure that the recommendations and areas of identified corrective action in the audit are followed up?   ☐ ☐ ☐

Chapter 20

Management review

Objective of the chapter

The objective of this chapter is to explain what a management review is. This chapter will provide you with the skills necessary to perform a management review and identify what level of management review is required to maintain a functional EMS. After finishing this chapter, you should be able to answer the following questions:

✓ What is a management review?

✓ Why is a management review important?

✓ What level of management review is necessary to meet the requirements of ISO 14001 and EMAS?

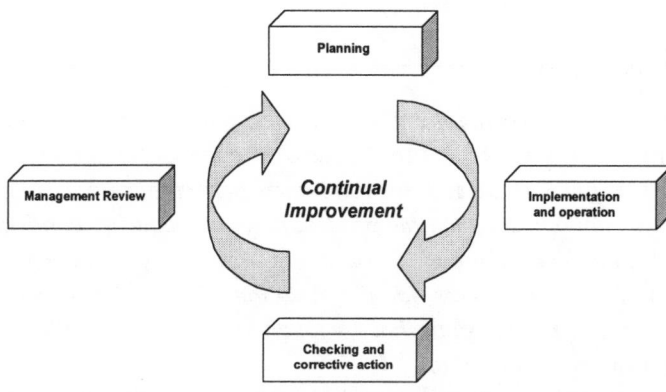

Introduction to the management review

Having audited your EMS, you will have undoubtedly found many areas of nonconformance or noticed where improvements to your environmental management system are needed. Logically, to improve the system and your environmental performance, the identified nonconformances must be corrected and improvements must be made where needed. This is the purpose of the management review. It is *the formal evaluation, by management, of the audit findings and the degree to which organization's environmental policy, objectives and targets and procedures are functioning as tools to improve environmental performance.*

Similarly, in the United Distillers case study, a management review would involve, among other things, a review of the findings and recommendations of the EMS audit to assess subsequently:

● The degree to which the environmental policy is being followed

● Whether the environmental objectives and targets are being realized

● The efficacy of the environmental management programmes

● The suitability of the entire EMS in light of changing circumstances, stakeholder views and the stated commitment to continual improvement

---

Hint

There are many variables that can change over time which directly affect the suitability of your EMS. Be on the lookout for changes to your products or activities, legislation, science and technology, market preferences, expectations of interested parties, reporting and communication.

Certification tip

In assessing your management review practices, your certifier will be looking for evidence that you are applying the process of continual improvement to your EMS such that gains in overall environmental performance are realized. For a smoother evaluation, explain the procedure you follow to conduct periodic reviews, identify what information you consider during your review, list who is involved in the review process, identify how the views of interested parties are taken into account and the steps that are taken to ensure the results of your management review are followed up.

Management reviews should be documented. These documents should be considered controlled documents and be referenced in your environmental management manual.

As with most of the EMS requirements discussed previously, the definitions of and requirements for management review differ slightly between ISO 14001 and EMAS. However, in both cases management review is effectively the same thing and is included in each of the EMS requirements for the same reason.

More precise requirements for certifiable management reviews are discussed in the next section, *Recommendations for successful EMS implementation.*

For examples of what other companies have done, visit the handbook web site at http://www.entropy-international.com/handbook/.

When you think you are finished, you can audit your performance and assess your success by using the check-list in the *Are you ready?* section at the end of this chapter.

Recommendations for successful EMS implementation

✓ You *must* conduct a management review to assess whether your EMS is adequate, appropriate and effective in relation to your organisation's overall intentions to improved corporate environmental performance.

✓ The management review of the EMS *must* be carried out by top management and by those who developed the EMS components being reviewed.

✓ The management review *must* be documented.

✓ The management review *must* assess the need to revise and improve the organisation's overall intentions with respect to the environment, including your environmental policy, objectives and targets, environmental management programmes and other key components of the EMS.

✓ The management review *must* assess the need to revise and improve the environmental policy, objectives and targets, environmental management programmes and other key components of the EMS in light of:

- EMS audits

- The degree to which the environmental policy has been followed

- The degree to which objectives and targets have been realized

- The efficacy of environmental management programmes and procedures

- Changing circumstances

- The views of stakeholders

- Your intention to improve continually your environmental performance

Are you ready?

	Yes	Partly	No
EMS Check-list **Management Review** *page 1 of 1*	■	◩	□

Have you conducted a management review to assess whether your EMS is adequate, appropriate and effective in relation to your organisation's overall intentions to improved corporate environmental performance?   ❑ ❑ ❑

Are your management reviews of the EMS carried out by top management and by those who developed the EMS components being reviewed?   ❑ ❑ ❑

Are your management reviews documented?   ❑ ❑ ❑

Do your management reviews assess the need to revise environmental policy, objectives and targets, environmental management programmes and other key components of the EMS?   ❑ ❑ ❑

Do your management reviews assess the need to revise and improve the environmental policy, objectives and targets, environmental management programmes and other key components of the EMS in light of:

- EMS audits?   ❑ ❑ ❑

- The degree to which the environmental policy has been followed?   ❑ ❑ ❑

- The degree to which objectives and targets have been realized?   ❑ ❑ ❑

- The efficacy of environmental management programmes and procedures?   ❑ ❑ ❑

- Changing circumstances?   ❑ ❑ ❑

- The views of stakeholders?   ❑ ❑ ❑

- Your intention to improve environmental performance continually?   ❑ ❑ ❑

Annex 1

Sample initial environmental review

This annex contains the Initial Environmental Review Report prepared for Dailuaine Distillery of United Distillers (now United Distillers and Vintners). This review was conducted in April 1997 by Entropy International. All information in this review is authentic but names have been changed to reflect the positions and responsibilities of the people identified.

As indicated in the Table of Contents, the original report was separated into tabs which for practical reasons have been omitted in this printing.

INITIAL ENVIRONMENTAL REVIEW

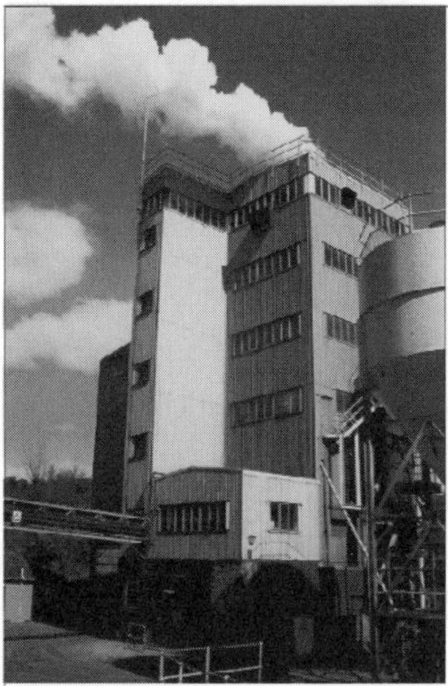

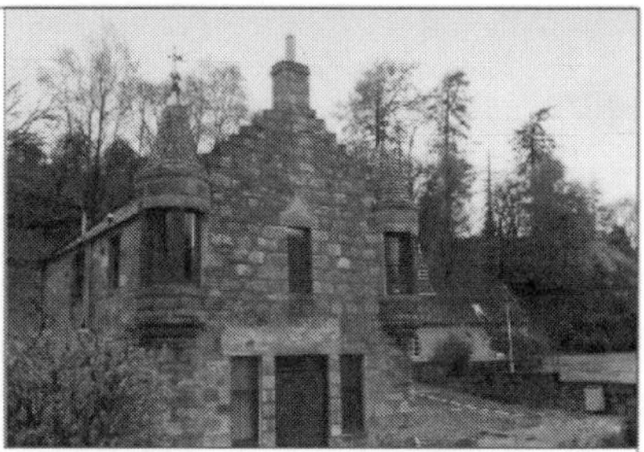

DAILUAINE
DISTILLERY

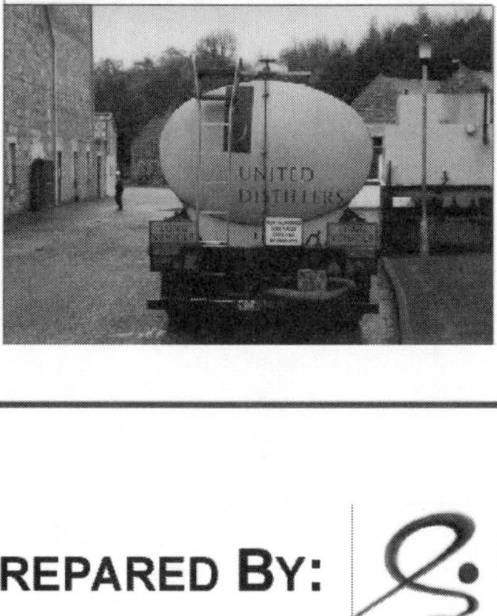

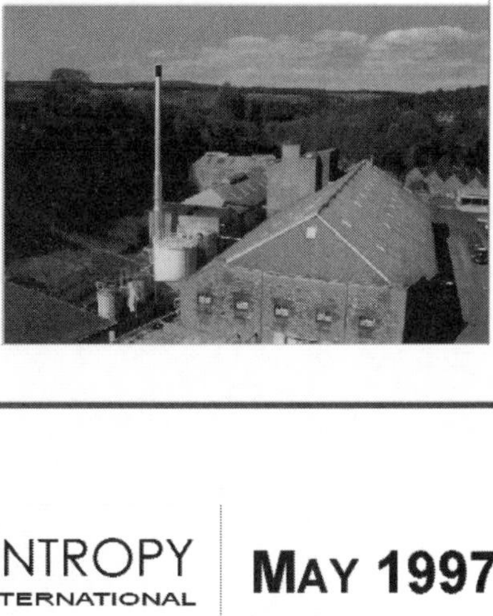

PREPARED BY:  **MAY 1997**

Table of contents

Table of contents (*continued*)

Executive summary

This initial environmental review has been conducted for United Distillers as part of their efforts to develop an environmental management system (EMS) certifiable to the ISO 14001 International EMS Standard. The initial review provides:

1   A review of existing environmental management practices at United Distillers' Dailuaine Distillery and recommendations for compliance with the requirements of ISO 14001 and Europe's EMAS regulation (Tabs 3 - 7).

2   Recommendations for a site environmental policy (Tab 4).

3   A description of the products from an environmental point of view (Tab 8) and identification of significant and notable environmental aspects and impacts (Tabs 9 - 16).

4   A methodology for continued identification of significant and notable environmental aspects and impacts and recommendations for improvements (Tabs 9 - 16 and 20).

5   A review of past environmental accidents and incidents (Tab 17).

6   A review of relevant environmental legislation, regulations and authorizations (Tab 18).

7   A methodology for linking the findings of the review to objectives and targets and an environmental management programme (Tab 20).

Most significant environmental aspects and impacts

- Water use: The plant uses a significant amount of water in the production process.

- Effluent to water: Large quantities of process waters are returned to the burns or sent for treatment to the bio-filtration plant for treatment and ultimate discharge into the Spey River.

- Energy use: Much of the process is energy intensive (particularly the dark grains plant). Use of heat exchange is exemplary.

- Use of chemicals (including special/hazardous) throughout the process.

- Use of radioactive substances in the evaporator and dark grains plant.

- Emissions to air: Flue gas emissions from the boilers and the dark grains plant, in addition to vented non-condensables during evaporation.

- Emissions to air: Exhaust from lorries delivering raw materials and effluent and picking up final products.

- Risk to worker health and safety in various stages throughout the entire process.

Main recommendations

✓ Update the current generic register of environmental hazards/effects in light of the environmental aspects and impacts identified in this review (see Tabs 9 - 16).

✓ Evaluate the environmental policy in light of the findings from this review. A site-specific policy should be developed (see Tab 3).

✓ Environmental objectives and targets should be formally decided on and set to reduce the aforementioned environmental impacts and improve overall corporate environmental performance (see Tab 4).

✓ A formal environmental management programme (action plan) should be developed and co-ordinated with ongoing environmental projects (see Tab 4).

✓ Develop a training programme for all employees in order to raise awareness of the EMS and communicate environmental policy, significant environmental aspects of the operations, regulatory responsibilities and the resulting objectives and targets for the site (see Tab 5).

✓ Update the current Malt Distilling EMS documentation such that the issues specific to the Dailuaine Distillery are reflected throughout.

✓ Integrate the EMS with the existing site management systems to realize the natural synergies between documentation, document control and the various management system components (see Tab 5).

✓ Consider the development of a 'Malt Distilling Intranet' or other electronic information system to facilitate cross-referencing of management system documentation and the dissemination of group reports and other relevant information.

✓ Develop the procedures suggested on the following page.

✓ Continually update and improve the EMS.

EMS procedures

✓ Procedures that are covered within the existing QMS and WH & S management system should be cross-referenced to the EMS and the generic Malt Distilling procedures should be reviewed and adapted to the operations of Dailuaine where appropriate. The following procedures should be adapted to satisfy the requirements of the EMS:

- A procedure for *identifying and controlling the environmental aspects and impacts* associated with your activities, functions, products and processes.

- A procedure for *identifying performance criteria* of all functions, activities, products and processes that have or could have a significant impact on the environment.

- A procedure for *identifying the training needs* required to implement, operate and maintain the EMS (see *Training, awareness and competence*, Tab 5).

- A procedure for *training personnel about the significant impacts of activities, functions, products and processes* (actual and possible) of their own work activities.

- A procedure for *identifying the likelihood of accidents and emergencies and the minimization, control and mitigation of environmental impacts* associated with such situations (see *Emergency preparedness and response*, Tab 5).

- A procedure for *internal communication* regarding environmental aspects, impacts and the EMS as a whole (see *Communication*, Tab 5).

- A procedure for *receiving, documenting and responding to external communication* associated with your organization's environmental performance or the EMS as a whole (see *Communication*, Tab 5).

- A procedure for *regularly monitoring and measuring* the functions, activities, products or processes that have or could have a significant impact on the environment (see *Monitoring and measuring*, Tab 6)

- A procedure dealing with *procurement and contract activities*, to ensure that suppliers and those acting on behalf of your company comply with your company's environmental policy.

- A procedure for the *approval of planned processes and equipment.*

- A procedure for *defining responsibility and assigning authority* for investigating the EMS for nonconformance with the stated requirements and rectifying nonconformances found (see *Nonconformance, correction and prevention*, Tab 6).

- A procedure for *identifying, maintaining and controlling environmental records* including monitoring and measuring results, training records, audit findings and review reports (see *Records*, Tab 6).

- A procedure for *controlling all related EMS documents* to ensure that: they can be located; they are reviewed, revised and approved regularly; current versions of documents are kept and available where needed; and obsolete documents are removed or otherwise marked accordingly (see *EMS documentation*, Tab 5).

- A procedure for *periodic EMS audits* (see *EMS audit*, Tab 6).

This initial environmental review was conducted and compiled May-June 1997 by Entropy International Environmental Consulting Limited and represents an objective third-party assessment of the operations at the United Distillers, Dailuaine Distillery.

Gary Robinson
Review Team Leader   14-6-97

1.   Introduction to the report

TAB 1

Purpose

The purpose of this initial environmental review was to provide the management and personnel at Dailuaine Malt Distillery with a snapshot of the present environmental state of affairs on
site.

The review is intended to identify the significant environmental aspects (causes) and environmental impacts (effects) of the activities, functions, products and processes at the site. The
review will provide management and personnel with a starting point towards developing an
environmental management system (ISO 14001 or EMAS compliant) to control and minimize
these environmental aspects and impacts and thus improve overall corporate environmental
performance.

Scope

This review covers the Dailuaine Malt Distillery in Aberlour, Scotland. The review includes the
main distillery site and the nearby wastewater treatment plant. The review addresses all steps
of the distillery process at Dailuaine but does not address the malting stage, which occurs at
separate geographic locations.

Methodology

The review process has included: a site visit to Dailuaine; examination of relevant site documents; interviews with site personnel and other stakeholders; and the collection of video, audio
and photographic information.

Project team

Gary Robinson, Review Team Leader

Hewitt Roberts

Time-frame of the review

The (on-site) review was conducted between April 22 and 25, 1997. The resulting report was
created and compiled May-June, 1997.

2.   Overview and general information

TAB 2

The company

Dailuaine Malt Distillery's main product is single malt whiskey spirit with the pot ale and draff by-products from the distilling process being converted to animal feeds as a secondary saleable product. The site at Dailuaine produces approximately 2.2 million litres of spirit annually, approximately 2,000 tonnes of pelletized animal feeds annually and employs 29 people on the site.

Dailuaine is owned by United Distillers which in turn is the spirits division of Guinness PLC. Guinness is likely (Dec 1997) to merge with Grand Metropolitan to form one of the world's largest food and beverage producers.

United Distillers see themselves as a very proactive company in relation to the environment and the distilling industry as a whole. United Distillers recognize that to ensure that their main raw materials (water and barley) remain wholesome and thus their products retain their present world class standards, they are unconditionally dependant on a clean environment and must consequently minimize the environmental impact of all of their operations.

United Distillers are members of the Malt Distillers Association of Scotland and encourage management at their sites to act as the 'environmental champions' of the EMS development.

The site

The site is two miles south-west of Aberlour in a quiet rural setting nestled in the Spey Valley.

The site is comprised of five main areas: the *distillery building* (including the mash tuns, wash backs, stills, boiler and fuel tanks); the *former maltings building* (including malt silos, mill room and grist cases, canteen, engineers' workshops, brewer's office, disused steeps, saladin boxes, kiln and flat bed barley store of the maltings); the *dark grains plant* (including pot ale evaporator, draff silos, dark grains dryer and pelletizer, outloading bins and fuel storage); the *warehouses* (including the filling store, barrel store); and the *effluent treatment plant* on the banks of the River Spey about $^1/_3$ mile downstream. The distillery and treatment plant occupy approximately 30 acres of land and, with the exception of the plume from the dark grains plant, the site is virtually hidden from the main road (A95).

The distillery owns 100 acres of farmland (the Carron Mains) adjoining the site and the site area appears to be well kept, clean, tidy and organized. Most buildings were in good repair and maintenance and site upkeep appeared to be an important part of regular activities.

Topography, hydrology and geography

The site is located at the confluence of the Balliemullich Burn, the Carron (Green) Burn and the Burn of Derrybeg – all tributaries to the River Spey approximately $^1/_3$ mile downstream.

There are no known aquifers or wells in the vicinity and Dailuaine has the sole right to use the water in the burns for their processes.

There are a number of farms and other distilleries up- and downstream of the site, all of whom appear to be aware of Dailuaine's activities. There are no known complaints from them but Dailuaine has had discharges to the Spey leading to complaints.

Location of the site in relation to risk receptors and surroundings

The site is located at the foot of the Dailuaine Terrace and with the exception of approximately 14 residential houses on the terrace there is little nearby human dwelling. The closest school is between 1 and 1.5 miles from the site with the site almost exclusively surrounded by farmland.

There are no known sites or areas of cultural or natural significance in the immediate surroundings with the exception that the Spey River is a premier salmon fishing river in Scotland and renowned world-wide for its fishing and natural beauty.

As Dailuaine is in a valley there are no significant prevailing winds that affect the site.

Other local industry

With the exception of a number of farms in the nearby vicinity and the aforementioned 100-acre farm at Carron Mains (leased from Dailuaine) there is no local industry of noteworthy mention.

Site history

Malt distilling has taken place at the site since 1851 and prior to that it is assumed to have been used as farming land. It appears unlikely that the site could have been seriously contaminated in the past or that the activities of days gone by would have any notable effect on present activities.

There have been two flash floods in recent years (1970 and 1982), with one causing the on-site reservoir to overflow and the second leading to greater flooding of the cooperage, resulting in damage to buildings and the site. A culvert was added to cope with future floods but the risk of flooding still remains.

There were two fatalities (1990) in the effluent plant as a result of asphyxiation from hydrogen sulphide in one of the sumps.

3.1 Commitment and policy

TAB 3

3.1.1 Overall management description

A) Gap analysis

The check-list over the page reviews the overall requirements for an EMS as stipulated by the International Standard ISO 14001:1996 and the EC's Eco-Management and Audit Scheme (EMAS), as well as the level of integration with existing management in relation to Dailuaine's current position.

B) Current practice

United Distillers' Dailuaine Distillery is aiming for certification according to the ISO 14001 environmental management system standard. Top management of the company recognizes the need for an EMS and has committed to providing support for the project.

Dailuaine has chosen to implement an EMS in order to formalize the existing environmental management practices such that they can be practised throughout Malt Distilling in a similar manner. By implementing a structured EMS, the company hopes to address environmental issues in a more efficient and transparent way. The EMS is seen as an integral and natural part of the already existing management system and as an important part of future business operations.

Development of Dailuaine's formalized EMS is being led by Wirral Green in co-ordination with Jim Nasium and Grant Fromage. The aim is ultimately to involve the whole workforce in the implementation and operation of the system. No dates have been set for implementation or certification at this time.

Existing environmental management practices are outlined in the Malt Distilling management system environmental manual. The system was developed in June 1995 and is generic for all of Malt Distilling and follows the requirements of BS 7750.

Dailuaine operates according to the ISO 9001 international quality standard and has a worker health and safety management system in place (HSG 70). The worker health and safety system is currently being updated.

The company culture can be characterized through United Distillers philosophy of continuous improvement in all spheres of its business activities.

<table>
<tr><td colspan="2">

EMS Check-list
Overall Management

page 1 of 1
</td><td>

■ ◩ ◻

Yes Partly No
</td></tr>
<tr><td colspan="2">

Has Dailuaine established and maintained an environmental management system (EMS) in accordance with the requirements outlined in ISO 14001:1996?
</td><td>◩</td></tr>
<tr><td colspan="2">

Has Dailuaine established and maintained an environmental management system (EMS) in accordance with the EC Council Regulation No. 1836/93 regarding EMAS?
</td><td>◩</td></tr>
<tr><td colspan="2">

Has Dailuaine established links between the quality management system (QMS) and the environmental management system (EMS)?
</td><td>◩</td></tr>
<tr><td colspan="2">

Has Dailuaine established links between the worker health and safety management system (WHS) and the environmental management system (EMS)?
</td><td>◩</td></tr>
</table>

C) Findings

✎ Many of the formal EMS components exist at the site. However the system documentation is not yet fully compliant with the requirements of ISO 14001 or EMAS.

✎ May of the existing EMS components have not been adapted to the specific activities and operations of the Dailuaine Distillery.

✎ Documentation, records and links between the QMS, HSG 70 and the EMS have not been formally established.

D) Recommendations

✓ Focus on the documentation and EMS recommendations, as outlined in this review, to achieve compliance with the international conditions for an EMS as outlined by ISO 14001 and EMAS.

✓ The existing overall Malt Distilling EMS components should be adapted to the specific activities and operations of the Dailuaine Distillery.

✓ Where possible links between the QMS, HSG 70 and EMS should be established and maintained in order to clarify the systems and avoid unnecessary documentation overlaps.

3.1.2   Environmental policy

A) Gap analysis
The check-list over the page assesses compliance with the minimum requirements for an environmental policy as stipulated by the International Standard ISO 14001 and by the EC Council Regulation No. 1836/93 regarding EMAS.

B) Current practice
United Distillers have an environmental policy and an environmental policy statement. There is also a brochure explaining the environmental policy. The existing policy is framed and displayed at the site although employees at the site are not sufficiently aware of its application to Dailuaine.

United Distillers have the following guiding aims:

● To know the consumer

● To build the business

● To lead the market wherever the company operates

United Distillers states that 'an enlightened commitment to its consumers and the communities in which it operates world-wide is an integral part of sound business practice'. It is stated that responsibility for the environment is important to the company as the company, more than other industries, relies on a healthy environment to provide water, cereals and other natural ingredients essential to the industry.

United Distillers claim to assess and measure environmental impacts carefully and to run programmes designed to conserve natural resources, reduce energy and waste, and wherever possible enhance the natural environment. Commitment to use of recyclable packaging, energy and raw material conservation, control of atmospheric pollution and responsible water management are other core environmental values.

Dailuaine is aiming to develop a site policy in order to address the environmental issues relevant to its operations.

EMS Checklist **Environmental Policy** *page 1 of 1*	 Yes Partly No
Does Dailuaine's environmental policy state their organization's principles and intentions in relation to their overall environmental performance?	■
Is Dailuaine's environmental policy appropriate to the nature, scale and environmental impacts of their activities, products and processes?	◪
Does Dailuaine's environmental policy include a commitment to continual improvement and the prevention of pollution based on an acceptable methodology?	■
Does Dailuaine's environmental policy include a commitment to comply with relevant environmental legislation, regulations and other requirements to which they subscribe?	■
Is Dailuaine's environmental policy documented, implemented, maintained, reviewed and made known to all employees?	◪
Does top management endorse Dailuaine's environmental policy?	■
Is Dailuaine's environmental policy available to the public?	◪
Does Dailuaine's environmental policy provide the framework for setting and reviewing environmental objectives and targets?	■
Is Dailuaine's environmental policy clear, concise and written in non-technical language interpretable by both internal and external parties?	■
Does Dailuaine's environmental policy include a commitment to the development and implementation of their EMS?	◪
Does Dailuaine's environmental policy include a commitment to the development of, and adherence to, corporate standards in the absence of legislation?	■
Does Dailuaine's environmental policy embody a life cycle approach to the environmental impacts of their activities, products and processes?	☐

C) Findings

✎ United Distillers environmental policy has been approved by top management.

✎ United Distillers environmental statement and environmental policy are applicable to the operations at Dailuaine.

✎ The environmental policy fulfils most of the requirements of the standard. However, the environmental policy has not yet been fully implemented, maintained and communicated.

✎ The environmental policy is available to the public although it is generic for all of United Distillers and therefore does not give a clear indication of practices specific to the operations of Dailuaine.

✎ The environmental policy includes no statements that are specific to the operations at Dailuaine and therefore does not provide a framework for the setting of realistic objectives and targets.

✎ The policy includes a commitment to the development and implementation of many components of an EMS but does not specifically indicate the adoption of a formal environmental management system to accomplish this.

✎ The environmental policy does not embody a life cycle approach to assessing the environmental impact of products and process.

D) Recommendations

✓ A site-specific environmental policy should be developed in light of the findings of the initial environmental review.

✓ A site-specific environmental policy should address environmental issues of particular relevance to the Dailuaine Distillery.

✓ The site environmental policy should form the basis for setting of objectives and targets.

✓ The environmental policy should be documented, implemented, maintained and communicated to all employees.

✓ The environmental policy should be approved by top management and signed by the Managing Director (Grant Fromage).

✓ The United Distillers environmental statement also works well for Dailuaine and should be adopted.

✓ The environmental policy should be made available to the public.

✓ Dailuaine's policy should make reference to the development and implementation of an environmental management system.

✓ The adoption of a life cycle approach to the environmental impacts of Dailuaine's activities, functions, products and processes should be considered for inclusion in the environmental policy.

Below is a suggestion for an environmental policy for Dailuaine. The suggested policy follows the methodology inherent to the United Distillers main policy while adding more specific statements that are applicable to the Dailuaine Distillery. Please note that this is only a guideline for a suitable site environmental policy and should not be entered into lightly, as the environmental policy is a crucial document for maintaining an effective EMS.

> *Dailuaine's environmental*
> *policy is shown on page 111*

3.2   Planning

TAB 4

3.2.1   Environmental aspects

A) Gap analysis

The check-list over the page assesses compliance with the minimum requirements for a register of environmental aspects and impacts (effects) as stipulated by the International Standard ISO 14001 and by the EC Council Regulation No. 1836/93 regarding EMAS.

B) Current practice

The site does not keep a formal register of environmental aspects and impacts relevant to the site. However, the *UMGD Malt Distilling Environmental Manual* includes an analysis of 'Environmental Hazards/Effects Under Normal and Abnormal Conditions'. This analysis was documented June 1995 and is generic to all the UMGD Malt Distilling sites.

See also the *Review of activities, products and processes* for a description of Dailuaine's on-site processes and significant environmental aspects.

C) Findings

✔ Major environmental aspects relevant to the company have been identified in the EMS manual (*UMGD Environmental Manual*, Section J, 'Environmental Hazards and Effects Under Normal and Abnormal Conditions'). This information was last updated June 1995 and is general to all of Malt Distilling.

✔ Operational procedures should cover all significant aspects and impacts as identified by this initial environmental review.

✔ Presently, there is no formal procedure for identifying environmental aspects and impacts and assessing their significance.

✔ Incidents, accidents and potential emergency situations are currently covered by the worker health and safety management system.

✔ The UMGD register of environmental hazards/effects has not been updated since June 1995, nor has a date for revision been identified.

EMS Check-list

Register of Environmental Aspects & Impacts

page 1 of 2

■	◪	☐
Yes	Partly	No

Has Dailuaine compiled a register of aspects and impacts identified as significant? — ◪

Does Dailuaine's register of aspects and impacts include (where significant):

- All inputs to their activities, products or processes? — ◪

- All outputs from their activities, products or processes? — ◪

- All air emissions (controlled and uncontrolled) from their activities, products or processes? — ◪

- All effluents (controlled and uncontrolled) from their activities, products or processes? — ◪

- The generation or disposal of solid and other waste (particularly hazardous waste) associated with their activities, products or processes? — ◪

- Any contamination of land as a result of their activities, products or processes? — ◪

- All uses of raw materials and natural resources associated with their activities, products or processes? — ◪

- All other discharges or emissions associated with their activities, products or processes (thermal energy, noise, odour, dust, vibrations and visual impact)? — ◪

- All environmental issues of local or community relevance associated with their activities, products or processes and any issues associated with the company and its environmental performance? — ◪

EMS Check-list
Register of Environmental Aspects & Impacts

■ ◪ ☐
Yes Partly No

page 2 of 2

Does Dailuaine's register of aspects and impacts identify the significant aspects and impacts arising from:

- All normal operating conditions?   ◪

- All normal activities, products and processes?   ◪

- All or any abnormal activities, products and processes?   ◪

- All accidents and potential emergency situations associated with their activities, products and processes?   ◪

- All past, present and planned activities, products and processes?   ◪

- The full life cycle of their products?   ☐

Is Dailuaine's register reviewed regularly and revised accordingly?   ☐

Is Dailuaine's register documented and presented in a clear, concise and easy-to-understand format?   ■

Does Dailuaine's register differentiate between direct and indirect aspects and impacts?   ☐

Does Dailuaine's register include the procedures employed to identify environmental aspects and impacts and their significance?   ☐

Does Dailuaine's register identify the procedures or instructions employed to assess significant environmental aspects and impacts?   ■

Is Dailuaine's register kept in the environment management manual?   ■

D) Recommendations

✓ The current generic register of environmental hazards/effects should be revised in light of the environmental aspects and impacts of the operations identified in this review.

✓ A procedure for identifying environmental aspects and impacts needs to be established.

✓ A procedure should be written for controlling all significant effects that aren't covered by the site environmental management procedures J1-J20. Subsequently, procedures J1-J20 should be reviewed and specifically adapted to Dailuaine's operations where possible.

✓ Opportunities for integrating the environmental management activities with the worker health and safety management system should be investigated.

3.2.2   Environmental legislation and regulations

A) Gap analysis
The check-list over the page assesses compliance with the minimum requirements for environmental legislation and regulations as stipulated by the International Standard ISO 14001 and by the EC Council Regulation No. 1836/93 regarding EMAS.

B) Current practice
Legal requirements are identified according to a procedure specified in the *UMGD Malt Distilling Environmental Manual*, Section H. Legislative documents are identified and listed in the same section of the manual.

See *Review of relevant environmental legislation, regulations, authorizations and industry codes of practice* for further detail.

Legislation, applicable to environmental matters and relevant to Malt Distilling, is identified by the Regulatory Group, ITS and the Environmental Manager (Wirral Green).

The supervising authorities for Dailuaine consist of:

1   Morray Council (water quality)

2   Scottish Environmental Protection Agency (SEPA)

United Distillers is also a member of the Scotch Whisky Association and Malt Distillers Association of Scotland.

C) Findings

✎   Dailuaine has a procedure for identifying and having access to legal requirements. The purpose of this procedure is to gain a sound understanding of all environmental legislation to ensure compliance with applicable regulations and legislation. This seems to be in accordance with standard requirements.

✎   To date there is no site-specific register of legislation that links relevant legislation and regulations to the significant environmental aspects and impacts.

EMS Check-list

Environmental Legislation and Regulations

page 1 of 2

■ ◨ □
Yes Partly No

Does Dailuaine have a written procedure for identifying and having access to legal requirements and other regulations that are applicable to Dailuaine?

■

Does Dailuaine comply with all the identified environmental legislation and regulations and with other requirements to which Dailuaine and United Distillers subscribe?

■

Has Dailuaine compiled a register of all environmental legislation and regulations associated with their identified aspects and corresponding significant environmental impacts, including other requirements to which Dailuaine and United Distillers subscribe?

◨

Is Dailuaine's register of environmental legislation and regulations kept up to date and revised when necessary?

■

Does the register identify the regulatory bodies associated with the identified legislation and regulations and briefly explain their activities and jurisdiction?

■

Does the register include reference to authorizations or permits required by Dailuaine?

□

Does the register include all site planning arrangements required for the site?

■

Does the register include all relevant regulations covering air emissions including quality and/or quantity restrictions?

■

Does the register include all relevant regulations covering discharges to water including quality and/or quantity restrictions?

■

Does the register include all relevant regulations covering water use?

■

EMS Check-list

Environmental Legislation and Regulations

page 2 of 2

Yes Partly No

Does the register include all relevant regulations covering waste disposal such as due care, packaging use and disposal or minimization strategies?

Does the register include all relevant regulations covering the use, storage and disposal of hazardous or special substances?

Does the register include all relevant regulations covering contaminated land or the potential for land contamination?

Does the register include all relevant regulations covering energy use, fuel use, land use and natural resource use?

Does the register include all relevant regulations covering discharges of thermal energy, noise, odour, dust, vibrations and visual impact?

Does the register include all relevant regulations covering aspects of worker health and safety that may also have environmental applicability?

Does the register include all relevant regulations covering the protection of biodiversity or impact on the local or global ecological environment?

Does the register directly link identified legislation or regulation to individual activities, products or processes or Dailuaine's activities?

D) Recommendations

✓ Create a site-specific register that directly links or makes reference between the significant environmental aspects and impacts and the corresponding legislation or regulations for that activity.

✓ Improve the procedure for identifying and having access to other guiding principles and controlling documents (including the environmental policy) relevant to the site.

✓ Everyone at the site should receive awareness training of the importance of complying with legislation and the possible consequences of noncompliance or contravening the law applicable to their place of work.

3.2.3 Objectives and targets

A) Gap analysis

The check-list over the page assesses compliance with the minimum requirements for environmental objectives and targets as stipulated by the international standard ISO 14001 and by the EC Council Regulation No. 1836/93 regarding EMAS.

B) Current practice

A generic set of environmental target areas identified for the plant can be found in the *UMGD Environmental Manual*, Section F. These cover:

- Energy use

- Water use

- Effluent produced

- Discharge consent compliance

- Dark grains recovery

- Pot ale syrup recovery

- Wet draff recovery

- Solid waste

- Gaseous emissions

It is the responsibility of the site's general manager to prioritize and set environmental targets. Compliance with legislation and the prevention of spillage (containment) were noted as site objectives.

Dailuaine has set environmental targets to the year 2000 in its Environmental Strategy Plan. These cover:

- Copper removal from spent lees (1996).

- Replace balancing and feed tanks with new round vessels complete with mixer pH control/level control (1997).

- Install cooling tower/air blast cooler or equivalent heat rejection system or pipe to River Spey (1997).

- Scrubber and WESP on DGP dryer (1997).

- Demolish existing concrete effluent tanks (1998).

EMS Check-list
Environmental Objectives and Targets

page 1 of 2

Yes   Partly   No

Have site objectives and targets been set to improve environmental performance?

Are Dailuaine's objectives overall performance goals?

Are the objectives reflected in Dailuaine's environmental policy?

Are Dailuaine's objectives detailed goals, in terms of environmental performance, which have been set at the site?

Are Dailuaine's objectives specific, realistic and achievable?

Are the objectives aimed at the continual (continuous) improvement of Dailuaine's environmental performance?

Are Dailuaine's objectives directly related to the significant environmental impacts of Dailuaine's activities, products or processes as determined by the findings of an IER?

Have the objectives been established, maintained and documented for all relevant activities, products and processes within Dailuaine?

Are the objectives consistent with the legislative and regulatory compliance requirements of the site?

Have the objectives been set considering the views of internal and external stakeholders as well as the financial, operational and organizational parameters of Dailuaine?

Have Dailuaine's objectives been implemented and are they regularly reviewed and revised, where necessary, with the endorsement of top management?

Are Dailuaine's objectives supported with sufficient human and financial resources required for their achievement?

EMS Check-list
Environmental Objectives and Targets

page 2 of 2

Yes Partly No

Are Dailuaine's targets detailed and quantifiable performance require-ments?

Have Dailuaine's targets been developed to meet the environmental objectives set?

Are the targets measurable with set dates against which progress can be measured?

Do Dailuaine's objectives and targets encompass a preventative approach to pollution wherever practicable?

Do Dailuaine's objectives and targets make use of methodologies such as the use of cleaner technology, BATNEEC or EVABAT, wherever possible?

Are Dailuaine's objectives and targets set out in an environmental management programme (action plan) specifying action steps to be taken, schedules, resources and responsibilities?

C) Findings

✐ The existing site targets are aligned mainly with respect to legislated conditions and have not been set with specific time-frames for completion (yearly figures).

✐ The target specifying the removal of copper from spent lees has not been met.

✐ The generic overall objectives for Malt Distilling are reflected in the United Distillers environmental policy but were not visibly apparent at the site level.

✐ The current environmental objectives are consistent with legislative and regulatory compliance but do not take into account the views of external stakeholders.

D) Recommendations

✓ Site-specific objectives and targets need to be set with time-frames and an identified percentage of reduction/improvement where possible.

✓ Site-specific targets need to be approved by the general manager and periodically reviewed and revised where necessary.

✓ Sufficient human and financial resources should be allocated to ensure stated targets are met within the scheduled time.

✓ Objectives and targets should be evaluated and developed according to the notable and significant environmental impacts as identified in Tabs 9 - 16, legislative requirements and the statements made in the environmental policy.

✓ Objectives should be developed that specifically address each significant and notable environmental aspect indicated in Tabs 9 - 16.

✓ The views of external stakeholders should be consulted and considered when establishing site environmental objectives and targets (see *Communication*, Tab 5).

3.2.4   Environmental management programmes

A) Gap analysis

The check-list over the page assesses compliance with the minimum requirements for an environmental management programme as stipulated by the International Standard ISO 14001 and by the EC Council Regulation No. 1836/93 regarding EMAS.

B) Current practice

The *UMGD Malt Distilling Environmental Manual* does not specify criteria for an environmental management programme.

Responsibility for implementing an environmental management programme is that of the general manager of the site. The co-ordination of an action plan at Dailuaine is informal and is currently focusing on dryer abatement technology, bio-plant improvements and heat recovery.

Other on-going or recently completed environmental projects at United Distillers, Dailuaine include:

- First environmental audit (1992)

- Upgrading of the bio-plant (1992)

- Second environmental audit (1995)

- Changing from heavy fuel oil to (dual fired) natural gas burners (1995)

- Control system for dark grains plant (1996)

- Copper removal system for the bio-plant (1997)

- Installation of two additional UV sterilization units (1997)

- Completion of this initial environmental review (1997)

- ISO 14001 certification preparation (1997)

- Participation in the 'InterAction for the Environment' project (1997)

EMS Check-list
Environmental Management Programme

page 1 of 1

	Yes	Partly	No
	■	◩	☐

	Yes	Partly	No
Has Dailuaine established and does it maintain environmental management programmes (environmental action plans) to meet the objectives and targets that have been set?		◩	
Are the environmental management programmes the recipe for achieving the objectives and targets that have been set?		◩	
Do the environmental management programmes have milestones, deadlines (time-frames) and assigned responsibilities?		◩	
Do the environmental management programmes establish the means and schedules for achieving objectives and targets and answer what, when, who, how and what next?		◩	
Are the environmental management programmes regularly reviewed (targets, budgets, responsibilities etc.) and revised accordingly?		◩	
Are the environmental management programmes revised in light of any new activities, products or processes?		◩	
Are Dailuaine's environmental management programmes documented and kept in the environmental management manual?			☐
Are Dailuaine's environmental management programmes the ultimate responsibility of upper management and have they been developed by those most closely associated with it?		◩	
Do the actions in the environmental management programmes have their own correlating objective, description, budget, evaluation procedure, start and finish dates, and training requirements?			☐
Are the methods of evaluation for the environmental management programmes documented and agreed upon?		◩	
Does this evaluation include how the programmes will be monitored, by whom, how problems and departures from your programmes will be dealt with and who is responsible for initiating and monitoring corrective action taken?			☐

C) Findings

✏ There are already several environmental projects operating within the company. However, as there is neither a formal action plan nor an overview of environmental projects and measures taken, it follows that there are no benchmarks for progress beyond the assessments made in the environmental audits of 1992 and 1995.

✏ Numerous recommendations were stated in the environmental audits (92 & 95) but evidence of formal management programmes to meet them was not forthcoming during this review.

✏ The current site environmental management programme is informal and not easy to ascertain without discussion with those most closely associated with it.

D) Recommendations

✓ Develop formalized environmental management programmes to achieve objectives and targets.

✓ The environmental management programmes should assign responsibilities for achieving the objectives and targets at the different levels of the organization.

✓ The environmental management programmes must include time-frames and quantifiable targets for each objective (see Environmental Action Plan Template, Annex 2, Tab 20).

✓ The environmental management programmes should be regularly reviewed with targets, budgets and responsibilities being revised accordingly.

✓ A procedure should be developed to ensure environmental management programmes are revised in light of any new activities, products or processes.

✓ The environmental management programmes must be documented and should be kept in the environmental management manual.

✓ The actions within the environmental management programmes should have their own description, purpose (objective), budget, evaluation procedure, start and finish dates and training requirements for those involved (where necessary).

✓ Methods of evaluation of the environmental management programmes should be documented and agreed upon. Evaluation should include how the programme will be monitored, by whom, how problems and departures from the programme will be dealt with and who is responsible for initiating and monitoring corrective action taken.

3.3-3.5   Other sections

(*TABS 5 - 7*)

The following sections have been omitted to save space:

   3.3   Implementation and operation

       3.3.1 Structure and responsibility

       3.3.2 Procedures

       3.3.3 Training, awareness and competence

       3.3.4 Communication

       3.3.5 EMS documentation

       3.3.6 Document control

       3.3.7 Operational control

       3.3.8 Emergency preparedness and response

   3.4   Measurement and evaluation

       3.4.1 Monitoring and measuring

       3.4.2 Nonconformance, correction and prevention

       3.4.3 Records

       3.4.4 EMS audit

   3.5   Review and improvement

In each case, the format of the report is the same as for the sections above, consisting of the following:

- Gap analysis
- EMS check-list
- Current practice
- Findings
- Recommendations

4.1 Overall site operation description

TAB 8

The overall operation at Dailuaine is the production of malt spirit for use in malt whisky. This process includes grinding malt, mixing it with hot water to extract the sugars from the barley starch, fermenting the solution to produce ethanol, distilling the ethanol to produce spirit, which is then packaged in oak casks and sent off-site to mature for a period of between 3 and 20 years and becomes whisky.

This process itself creates a number of waste streams that are further processed on site to produce 'dark grain' pellets which are sold to the farming community for cattle feed.

The main inputs to the process are water and malted barley. Dailuaine's main suppliers are essentially themselves as they buy malted barley from their own maltings. Dailuaine's main customers are again essentially themselves as the casked spirit is sold to United Distillers (UD) Commercial Division where it is either packaged as a UD 'single malt' product or sent to other distillers in a reciprocal arrangement for blending into a 'blended' whisky.

Overall Process Flow Chart

United Distillers, Dailuaine Malt Distillery

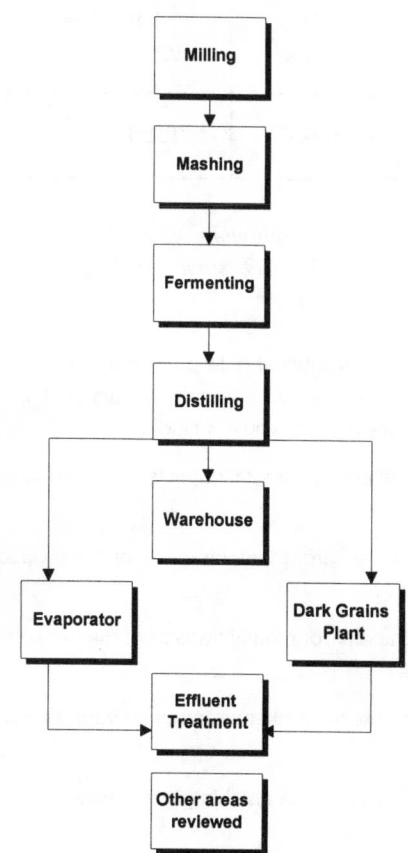

4.2 Description of the main products from an environmental point of view

A) Current practice

As mentioned above the main product of the site is malt spirit with the production of an animal feed by-product. The main inputs to the process are malted barley, water and energy.

According to the flow rates for a *medium-sized distillery*, Dailuaine Malt Distillery would use and produce approximately the following annually:

Input		Output	
Barley	6,240 tonnes	Water vapour	8,164 tonnes
Steep water	31,200 tonnes	CO_2	2,028 tonnes
Cooling water	624,000 tonnes	Cooling water	624,000 tonnes
Process water	43, 420 tonnes	Sludge	1,404 tonnes
Yeast	104 tonnes	Animal feed	2,236 tonnes
Effluent	63,492 tonnes	Whisky	1,664 tonnes
Total	**704,964 tonnes**	**Total**	**702,988 tonnes**

The difference between the input and output results above could be explained by not accounting for outputs such as NO_x, SO_x, CO, stones and other debris. Dailuaine's figures are approximately 125% of the figures above.

A full life cycle analysis of the product should involve a 'cradle-to-grave' investigation of the product in an attempt to reduce the environmental impact of all activities, functions and products associated with this process. This would include:

- An investigation into the environmental impacts of the production and use of barley, peat, water use and energy use.

- An investigation into the environmental impacts of the production of malt spirit (this initial environmental review).

- An investigation into the environmental impacts of the transportation and packaging associated with the product.

- An investigation into the social and environmental impacts associated with the use of their products.

- An investigation into the environmental impacts associated with the final disposal of their products.

To date, United Distillers has not conducted a formal life cycle analysis (LCA) of its products from an environmental perspective but a number of issues associated 'upstream and down-stream' of the malt spirit product are of recognized concern.

Having said that, many of the broader-reaching environmental issues (beyond the on-site process) are discussed in the United Distillers 'Issues' booklet produced by the Public Affairs Department, July 1992. The book discusses the following environmental considerations:

- General environmental policy and statement on the environment

- The use of peat and its extraction as a raw material

- Air pollution (site legislation focus, and smell abatement)

- Scottish forestry and its potential threats to water supplies

- Nuclear dumping (implications of waste disposal sites in Scotland)

- Water pollution (co-operation for future water quality standards)

- Packaging and developments to minimize the impacts

- Ingredient listing to safeguard consumer interests

- Production levels and forecasting demand

- Social aspects of alcohol

The importance of a life cycle approach is reiterated within the Guinness PLC Environmental Report. The Chairman's Letter states:

> 'Guinness PLC is dependent upon pure water, clean air and fertile soils for the production of its internationally renowned spirits and beers. It goes without saying that care for and protection of the environment must be a continuing priority throughout the business. We are determined to respond to this responsibility with a programme of well-directed and pro-active environmental management.'

B) Findings

✎ A formal LCA of United Distillers operations has not been conducted. There is at present no substantial quantification of the organization's overall environmental impact. However many of the major issues have been raised and discussed.

✎ The farming of barley and the use of pesticides, insecticides and fertilizers all have serious environmental implications. This issue has not been addressed.

✎ The 'Issues' booklet has not been updated within the last five years.

C) Recommendations

✓ Dailuaine and United Distillers should assess the environmental impacts of farming barley. Issues that should be addressed include the use of pesticides, insecticides and fertilizers (all have serious environmental implications).

✓ Dailuaine and United Distillers should investigate the possibility of purchasing, and encouraging the production of, 'organically-grown barley' and employment of sustainable farming practices.

✓ Dailuaine and United Distillers should investigate the possibility of using energy sources produced from sustainable means and develop a clear policy for the use of electricity produced from non-renewable and nuclear sources.

✓ Dailuaine and United Distillers should assess the environmental impacts of all the transportation involved in the production, distribution and use of their products and develop a strategy to minimize the impact associated with transportation.

✓ Dailuaine and United Distillers should further assess the environmental impacts of the packaging used for their products. An assessment of lightweight, recyclable, returnable or environmentally benign packaging should be considered. With recent EC legislation on packaging, this should be seen as a priority area.

✓ Dailuaine and United Distillers should assess the social and environmental impacts of the use of their products. A strategy should be developed to address further the issues of alcohol and health, alcohol and youth, alcohol and pregnancy, alcohol abuse, drinking and driving, and United Distillers should act proactively to minimize the impact of their product and promote socially responsible consumption patterns.

✓ Dailuaine and United Distillers should adopt a strategy to reduce the environmental impact associated with the final disposal of their products (packaging).

✓ The LCA approach to environmental impacts should be integrated into the existing Malt Distilling EMS.

4.3 Description of the main processes of the site

4.3.1 Milling

The malt distilling process begins with production of malt. In this process barley is dried, dressed of any field debris such as stones and straw and 'steeped' with water to start the germination process and promote enzyme activity. The 'malted' barley is then dried and ready for milling. No malting occurs at Dailuaine so 'malted barley' or 'malt' is delivered to the site by tipper lorry and stored for use.

In the milling process, the barley husk is kept with the barley flour to facilitate drainage later in the mashing process. The grist as it is then called is then ready for the mashing process. Prior to milling any contaminating material, such as stones or field debris, is dressed out to protect the milling equipment and prevent sparks from igniting the dust in the process. Dailuaine presently uses over 300 lorry loads of malt (7,300 tonnes) annually.

4.3.2 Mashing and fermentation

The grist is then placed in the mash tun where it is mixed with three batches of hot water to extract the grain sugars from the starch in the grist. During this rinsing the enzymes continue to break down the grain starches, producing sugar, and the sugary solution, 'worts', is drained off and sent to the fermentation process leaving behind cereal residues, know as 'draff'.

The worts are then cooled and yeast (3,650 Kg/week) is added and the batch is fermented for a period of approximately 4 days. After the fermentation is complete and all the sugar is converted to ethanol, the 'wash', as it is then called, is passed forward to the distillation process.

4.3.3 Distillation

In malt distilling, distillation is a two-stage process that removes the volatile components of the ethanol to produce alcohol 'spirit'. The first stage recovers the ethanol and the second stage refines the first, producing three output streams: the 'foreshots', the 'feints' and the spirit. The feints are returned to the first distillation stage and spirit is then piped to the warehouse.

4.3.4 Warehouse

In the warehouse the spirit is diluted with water from 68% to 63%, stored in oak casks and sent to off-site warehouses where it matures for up to 20 years. Dailuaine fills and ships approximately 400 casks weekly. This process is where the casks are prepared for filling which includes painting the cask ends, affixing identification bar codes and making minor repairs to the casks if necessary.

4.3.5   Evaporation

The first distillation residue, 'pot ale' as it is now called, is evaporated to remove water content and produce a 'pot ale syrup' solution with 46%-50% total solids. The syrup is then piped to the dark grains plant where it is added to the dried 'draff' to form pelletized animal feed.

4.3.6   Dark grains plant

At the dark grains plant the draff from the mashing process is mechanically squeezed to reduce its moisture content before it is dried with hot air and pelletized for use as animal feed. This process caters for all waste material from the distilling process (pot ale and draff) and the dark grains plant at Dailuaine distillery process their own pot ale and draff as well as from six to eight other nearby distilleries.

4.3.7   Effluent treatment plant

The wastewaters from the distillation that are not processed at the dark grains plant are collected and treated biologically at the nearby 'bio-plant'. Because the waste streams themselves are derived from evaporation or distillation they are generally deficient of nutrients such as nitrogen or phosphorus and are acidic. Nutrients are added to the waste stream to aid the biological treatment process.

The treatment process starts with a high rate bio-filtration stage that removes 90% of the polluting load and any surplus biomass from the effluent. These wastes are then settled, removed as sludge and disposed to grassland as a fertilizer. A second low rate bio-filtration stage removes the remaining organic matter and the final effluent is then discharged to the Spey River.

4.4 Milling

TAB 9

Milling Process Flow Chart

United Distillers, Dailuaine Malt Distillery

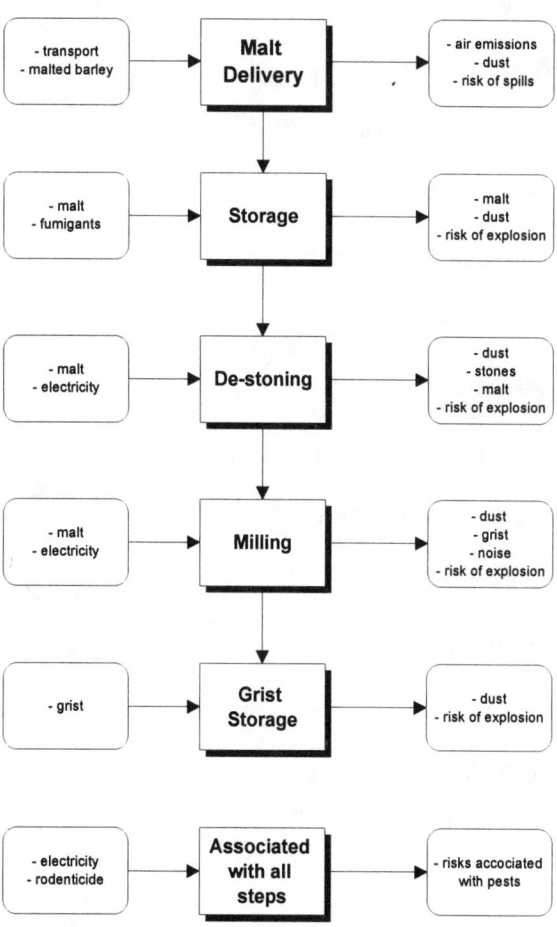

Milling: Significant (🌑*) and notable (↶) aspects and impacts

Note: Aspects and impacts are listed with *all significant aspects and impacts* appearing first (and in order of occurrence) followed by *all notable aspects and impacts* (in order of occurrence). This compiled list identifies the significance of environmental aspects and impacts based on their probability of occurrence and severity of environmental impact. It does not assess the level of control exhibited by the site.

🌑*   Transport emissions during malt delivery

🌑*   Use of fumigants for storage

🌑*   Dust during de-stoning

🌑*   Risk of explosion during de-stoning

🌑*   Noise during the milling stage

🌑*   Risk of explosion during the milling stage

🌑*   Dust during grist storage

🌑*   Risk of explosion during grist storage

🌑*   Use of small quantities of rodenticide

↶   Fuel for transport during malt delivery

↶   Use of barley in the process

↶   Dust during malt delivery

↶   Dust during storage

↶   Risk of explosion during storage

Milling: Findings

Water use

✏ Water use in this process is negligible.

Raw material use

✏ The main raw material in this process is malted barley. Dailuaine Distillery uses approximately 7,300 tonnes of malt annually.

Energy use

✏ Electrical energy is used for the de-stoning and milling equipment and for lighting.

Chemical use

✏ The process is occasionally fumigated with Methyl Bromide or Phosphine pH3 to guard against insect infestations.

✏ The rodenticides on site (warfarin – an acatonly-benzyl hydroxycoumarin compound) are supplied on a contract basis by Rentokil.

Effluents discharges

✏ There are no effluents discharged in this process.

Emissions to air

✏ Barley dust is prominent in all steps of the process after malt delivery.

✏ Phosphine (used in fumigation) has an odour characterized as 'evil-smelling'.

✏ As malt is trucked to Dailuaine, vehicle emissions are associated with this process.

Solid waste

✏ Stones from de-stoning process.

✏ Occasional waste barley from tipper spills.

Hazardous, special or restricted substance use

✏ Apart from the fumigants and rodenticides mentioned above there are no hazardous, special or restricted substances involved with this process.

Storage

✏ Barley is stored for milling.

Abnormal situations

✐ As this process is very dusty, there exists the risk of explosion in the presence of open flame or sparks.

✐ Barley spills encourages the presence of pigeons which brings with it health risks from droppings, carcasses etc.

Other

✐ Noise levels can exceed 85 dB A in the milling room.

Milling: Recommendations

✓ The use of Methyl Bromide and Phosphine should be reconsidered. They are both on the US EPA Toxic Release Inventory list of toxic substances and for which use must be reported annually. Similar restrictions should be expected in Europe and a proactive position to toxic use reduction should be adopted.

✓ Dailuaine should extend their environmental principles and policies to all of their site contractors and thus where possible impart a proactive and preventative approach to Rentokil.

✓ Methods for reducing emission of dust to the process should be assessed.

✓ Barley spills should be avoided and when they do occur should be cleaned up promptly.

Form 1

Process activity descriptions and environmental aspects

Site: Dailuaine Distillery **Date:** May, 1997

Main process: Milling

Individual process steps	Reference	Description of individual process steps	Aspects	
			Normal conditions	Abnormal conditions
Malt delivery		This process starts with the delivery to the site of malted barley. The barley arrives by lorry in 24 tonne loads and is poured into a ground-level holding area where it is stored for later use in the process.	Use of raw material (malted barley); transport emission; dust.	Malt spillage; health risks associated with pigeons attracted to spilled barley.
Storage		The barley is stored in a hloding area where it awaits delivery to the de-stoning milling process. The storage is fumigated annually to prevent accumulations of pests, insects and other contaminants.	Dust; fumigation with Methalin Bromide or Phosphine pH3 (very small amounts - 1 litre/bin/yr); storage of grist.	Solid waste to bin if spillage of malt; risk of explosion.
De-stoning		As this process is very dusty, there is an increased risk of explosion from sparks and barley dust. To minimize this risk the barley is de-stoned and passed through a magnetizer to remove any remaining debris.	Dust; stones as solid waste from process (very little - 50 Kg/wk); use of electricity.	Risk of explosion.
Milling		The dressed malt is then milled to a desired consistency. The grist as it is then called contains the barley flour and the husks. The husks aid in the following mashing process.	Dust; noise; use of electricity.	Risk of explosion.
Grist storage		After milling, the grist is stored for use in the next mashing process.	Dust.	Risk of explosion.
All process steps		Birds, rodents and insects are attracted to all steps of this process and preventative measures taken to minimize contamination.	Electricity for lighting; use of Warfarin rodenticide for rats (very small quantities).	

Form 2
Process activities & environmental aspects matrix (page 1)

Site: Dailuaine Distillery   **Date:** May, 1997

Process: Milling

Aspect reference numbers	General aspects	1	2	3	4	5	6
WU	**Water Use**						
WU01	Use of water from municipal sources						
WU02	Use of water from surrounding water courses						
WU03	Other water use						
EU	**Energy Use**						
EU01	Use of natural gas (not including use for transportation)						
EU02	Use of oil (not including use for transportation)						
EU03	Use of coal (not including use for transportation)						
EU04	Use of other fossil fuels (not including use for transportation)						
EU05	Use of fuel for transportation	X					
EU06	Use of energy from nuclear-generated sources						
EU07	Use of energy from hydro-generated sources						
EU08	Use of energy from wind-generated sources						
EU09	Use of energy from solar-generated sources						
EU10	Use of electricity from mixed sources			X	X		X
EU11	Other energy use						
CU	**Chemicals Use**						
CU01	Use of restricted chemicals		X				
CU02	Use of acidic chemicals (not on list of restricted chemicals)						
CU03	Use of basic chemicals (not on list of restricted chemicals)						
CU04	Use of solvents (not on list of restricted chemicals)						
CU05	Use of hydraulic oils, lubricants etc.						
CU06	Other chemical use						X
RU	**Raw Materials Use**						
RU01	Use of raw materials (hazardous, special or restricted)						
RU02	Use of raw materials (not hazardous, special or restricted)	X					

Form 2
Process activities & environmental aspects matrix (page 2)

Site: Dailuaine Distillery

Date: May, 1997

Process: Milling

Aspect reference numbers	General aspects	Process steps 1 2 3 4 5 6
RU	**Raw Materials Use (continued)**	
RU03	Use of packaging material (not including RU01 or RU02)	
RU04	Use of office materials (not including RU01, 02, or 03)	
RU05	Use of construction materials (not including RU01, 02, 03 or 04)	
RU06	Other raw material use (not including RU01, 02, 03, 04, or 05)	
ST	**Storage on Site**	
ST01	Storage of chemicals	
ST02	Storage of raw materials	X
ST03	Storage of hazardous, restricted or special substances	
ST04	Storage of waste (not hazardous, restricted or special)	
ST05	Storage of hazardous, restricted or special waste	
ST06	Other storage	
EW	**Effluents to Water**	
EW01	Discharge of effluent to treatment facility	
EW02	Controlled discharge of treated effluent to water course	
EW03	Controlled discharge of untreated effluent to water courses	
EW04	Uncontrolled discharge of treated effluent to water courses	
EW05	Uncontrolled discharge of untreated effluent to water courses	
EW06	Discharge of hazardous, restricted or special effluent	
EW07	Other discharges	
EA	**Emissions to Air**	
EA01	Emission of process gases/heat within the process (not up flue)	
EA02	Emission of flue gases/heat (not including NO_x, SO_x, particulates)	
EA03	Emission of NO_x	
EA04	Emission of SO_x	

Form 2
Process activities & environmental aspects matrix (page 3)

Site: **Date:**

Process:

Aspect reference numbers	General aspects	1	2	3	4	5	6
EA	**Emissions to Air (continued)**						
EA05	Emission of CO_2						
EA06	Emission of particulate matter (fly ash)						
EA07	Emission of dust or raw materials from within the process	X	X	X	X	X	
EA08	Emission of VOCs						
EA09	Emission of hazardous, restricted, special substances (not VOCs)						
EA10	Emissions from transport	X					
EA11	Other emission						
DL	**Disposal to Land**						
DL01	Disposal to municipal landfill			X			
DL02	Disposal to site landfill						
DL03	Disposal to incineration						
DL04	Disposal to recycling, reclamation or re-use						
DL05	Disposal of hazardous, restricted or special substances						
DL06	Previous soil contamination (actual and potential for)						
DL07	Other disposal						
OT	Other						
OT01	Vibrations						
OT02	Noise, smell				X		
OT03	Visual impact (include lights)						
OT04	Other						
AB	Risk of Abnormal Activity						
AB01	Risk of fire or explosion		X	X	X	X	
AB02	Risk of spillage, leakage or uncontrolled discharge	X					
AB03	Risk of spill etc. of hazardous, restricted or special substances						
AB04	Risk to worker health and safety						
AB05	Other abnormalities	X					

Form 3

Process environmental impact description and significance matrix

Site: Dailuaine Distillery

Date: May, 1997

Main process: Milling

Process steps	Aspect or impact identified	Ref. No.	Impact Description	Direct or indirect	Impact rating	Severity rating	Significance factor
Malt delivery	Fuel for transport	EU05	Depletion of non-renewable fossil fuel resources. Global oil resources estimated to be 70-80 yrs. combustion leads to emission of VOCs, NO_x, SO_x, CO_2 and thus air pollution, acidification, greenhouse gas production and global warming.	Direct	3	3	9
Malt delivery	Use of barley	RU02	Use of water, pesticides, insecticides and fertilizers during farming leading to land degradation, eutrophication and persistent accumulation of chemicals in the environment.	Direct	3	2	6
Malt delivery	Dust	EA07	Contribute to air pollution directly and indirectly as synergists or carriers of other pollutants. Can affect health and local environment.	Direct	3	3	9
Malt delivery	Transport emissions	EA11	Emission of VOCs, NO_x, SO_x, CO_2 and particulate matter to atmosphere and thus contribute to ozone depletion, air pollution, greenhouse gas production and global warming.	Direct	4	3	12
Malt delivery	Risk of spills	AB02	Impact can affect local environment and water ecology, impact on groundwater quality and risk to WH & S.	Direct	2	3	6
Malt delivery	Contamination from birds, rodents, insects	AB06	Risk of minor contamination of the product, poor housekeeping and associated risks to WH & S.	Direct	2	2	4

Form 3

Process environmental impact description and significance matrix

Site: Dailuaine Distillery **Date:** May, 1997

Main process: Milling

Process steps	Aspect or impact identified	Ref. No.	Impact Description	Direct or indirect	Impact rating	Severity rating	Significance factor
Storage	Use of fumigants	CU01	Highly poisonous liquid, when volatized, is capable of destroying vermin, insects, bacteria and moulds. Can act as a disinfectant.	Direct	3	4	12
Storage	Storage of barley	ST02	Negligible risk or impact.	Direct	2	1	2
Storage	Dust	EA07	Contribute to air pollution directly and indirectly as synergists or carriers of other pollutants. Can affect health and local environment.	Direct	3	3	9
Storage	Risk of explosion	AB01	Risk of explosions or fire from accidents or abnormal conditions. Could seriously affect WH & S and local environment.	Direct	3	3.5	10.5
De-stoning	Dust	EA07	Contribute to air pollution directly and indirectly as synergists or carriers of other pollutants. Can affect health and local environment.	Direct	5	3	15
De-stoning	Disposal of debris	DL01	Land use, degradation and contamination. Contribution to acidic landfill leachate high in Biological Oxygen Demand (BOD), ammonia, organic nitrogen, volatile fatty acids and other toxins. Affects local ecology and environment and is malodorous.	Direct	2	2	4
De-stoning	Risk of explosion	AB01	Risk of explosions or fire from accidents or abnormal conditions. Could seriously affect WH & S and local environment.	Direct	3	3.5	10.5

Form 3

Process environmental impact description and significance matrix

Site: Dailuaine Distillery   **Date:** May, 1997

Main process: Milling

Process steps	Aspect or impact identified	Ref. No.	Impact Description	Direct or indirect	Impact rating	Severity rating	Significance factor
Milling	Dust	EA07	Contribute to air pollution directly and indirectly as synergists or carriers of other pollutants. Can affect health and local environment.	Direct	3	4	12
Milling	Noise	OT02	Excessive or prolonged exposure to noise (typically more than 8 hrs above 85-90 decibels - average factory is 78 dB) leads to hearing loss. Noise pollution is the most common occupational hazard and can affect local ecology and WH & S.	Direct	4	3	12
Milling	Risk of explosion	AB01	Risk of explosions or fire from accidents or abnormal conditions. Could seriously affect WH & S and local environment.	Direct	3	5	15
Grist storage	Dust	EA07	Contribute to air pollution directly and indirectly as synergists or carriers of other pollutants. Can affect health and local environment.	Direct	3	4	12
Grist storage	Risk of explosion	AB01	Risk of explosions or fire from accidents or abnormal conditions. Could seriously affect WH & S and local environment.	Direct	3	5	15
All process steps	Use of electricity from mixed sources	EU11	Depletion of non-renewable natural resources. If source is nuclear power, impact increases through power plant water and raw material use (uranium 235, plutonium 239) and disposal of radioactive waste.	Direct	3	2	6
All process steps	Use of rodenticide	CU06	Highly poisonous substance for warm blooded mammals, risks to WH & S	Direct	3	4	12

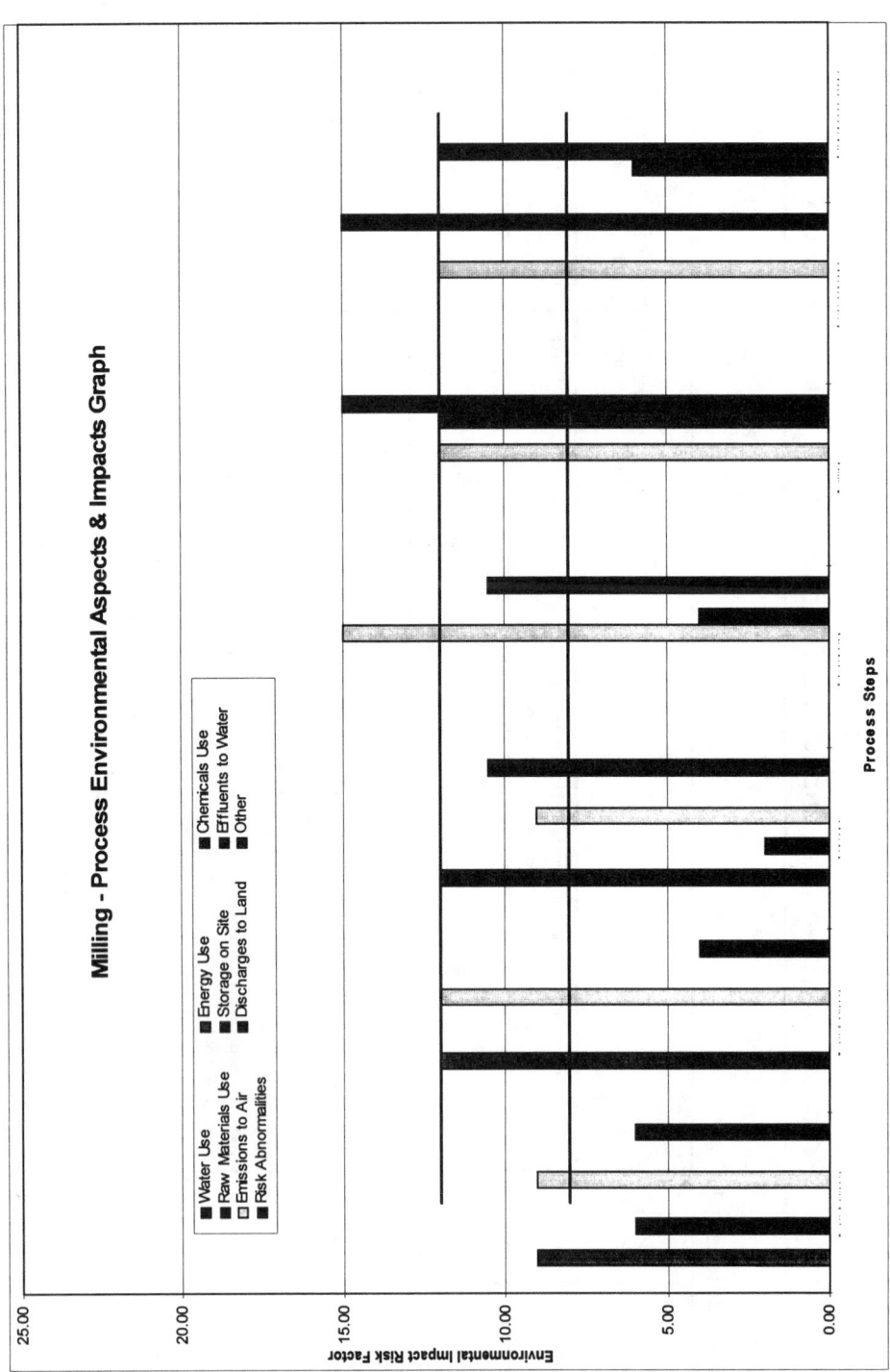

Milling - Process Environmental Aspects & Impacts Graph

4.4-4.11   Other processes

TABS 10-16

The following sections have been omitted to save space:

    4.5   Mashing and fermentation

    4.6   Distillation

    4.7   Warehouse

    4.8   Evaporation

    4.9   Dark grains plant

    4.10   Effluent treatment

    4.11   Other site activities

In each case, the format of the report is the same as for *Milling* above, consisting of the following:

- Process flowchart
- Significant and notable aspects and impacts
- Findings
- Recommendations
- Forms 1, 2 and 3
- Process environmental aspects and impacts graph

5. Review of past environmental accidents and incidents

TAB 17

While the past history of the site at Dailuaine seems clear and without mystique, the known history of the site only dates back to the middle 1800s. However, while it would appear that the risk of previous contamination is low, it is assumed that the site was predominantly used as farmland and thus there does exist the possibility that pesticides and other toxic farm residues may be present in the soil.

The following list describes the recent history of accidents and incidents at Dailuaine:

- In 1996 the cleaning out of distillery dam led to the release of organic silt into a small stream causing rapid depletion of oxygen, high levels of sulphide, fish deaths and complaints to SEPA from downstream fishing interests.

- In 1995 there was an accidental discharge of raw pot ale to surrounding watercourses through a mix up with valves in the evaporation process.

- There were two deaths in 1990 from hydrogen sulphide asphyxiation in the treatment plant sump house.

- In 1983 there was an incident where the driver of an oil tanker delivering oil dumped excess heavy fuel oil into the burn that flows to the Spey. A clean-up operation was required.

- Since the 1970s there have been three flash floods at the distillery and the threat of flooding still exists.

Like many accidents and incidents, many if not most of those associated with the operations at Dailuaine could have been prevented and although there is definitely a far greater level of overall awareness and prevention today, attention to minimizing risk and pollution prevention should be part of on-going site activities.

6. Review of relevant environment legislation, regulation, authorizations and industry codes of practice

TAB 18

A review of the environmental legislation, regulations, authorizations and industry codes of practices applicable to Dailuaine indicates that the site maintains an exemplary degree of organization, competence, compliance and record keeping in this area.

A full assessment of pertinent legislation applicable to Dailuaine was conducted in February 1997.

The following lists outline the legislation of importance identified.

Waste

- Control of Pollution (Amendment) Act 1989

- Environmental Protection Act 1990 (EPA 90) – Part II

- Controlled Waste (Registration of Carriers and Seizure of Vehicles) Regulation 1991 (SI 1991/1624)

- Controlled Waste Regulation 1992 (SI 1992/588), amended by SI 1993/566

- Environmental Protection (Duty of Care Regulations 1991) SI 1991/2839

- Special Waste Regulations 1996 (SI 1996/972)

- Waste Management Licensing Regulations 1994 (SI 1994/1056) as amended by SI 1995/288 and SI 1995/1950

- EU Regulation 259/93 on the Supervision and Control of Shipments of Waste within, into and out of the European Community

- The Carriage of Dangerous Goods (Classification, Packaging and Labelling) and Use of Transportable Pressure Receptacles Regulations 1996 (CPL2)

- The Carriage of Dangerous Goods by Road Regulations 1996 (CDG Road)

- Environment Act 1995 – Part V

- Landfill Tax Regulations 1996 (SI 1996/1527)

- The Landfill Tax (Quantifying Material) Order 1996 (SI 1996/1528)

- The Landfill Tax (Contaminated Land) Order 1996 (SI 1996/1529)

- Draft Producers Responsibility Obligation (Packaging Waste) Regulations

Air pollution

- Environmental Protection Act 1990 (EPA) – Part 1

- Environment Protection (Prescribes Processes and Substances) Regulations 1991, as amended by SI 1992/614, SI 1993/1746, SI 1993/2405, SI 1994/1392 and SI 1995/3247.

- Clean Air Act 1993 (CAA 93)

- Dark Smoke (Permitted Periods) Regulations 1958 (SI 1958/498)

- Clean Air (Height of Chimneys – Exemption) Regulations 1969 (SI 1969/411)

- Clean Air (Arrestment Plant – Exemption) Regulations 1969 (SI 1969/1262)

- Clean Air (Emission of Dark Smoke – Exemption) Regulations 1969 (SI 1969/1263)

- Clean Air (Emission of Grit and Dust from Furnaces) Regulations 1969 (SI 1971/162)

- Control of Asbestos in the Air Regulations 1990 (SI 1990/556)

- Environment Act 1995 (EA 95) – Part IV

- Air Quality Standards Regulations 1989 (SI 1989/317) Amended by SI 1995/3146

- EU Regulation 594/91 on Substances that Deplete the Ozone Layer as amended by 3952/92

- EU Regulation 3093/94 on Substances that Deplete the Ozone Layer

Water pollution

- Water Resources Act 1991 (WRA 91) – Part III

- Environment Act (Schedule 22)

- Surface Water (Dangerous Substances – Classification) Regulations 1989 (SI 1989/2286) as amended by SI 1992/337

- Surface Water (River Ecosystem – Classification) Regulations 1994 (SI 1994/1057)

- Environmental Protection Act 1990 – Part I – IPC

- Water Industry Act 1991 (WAI 1991) Part IV

- Trade Effluents (Prescribed Processes and Substances) Regulations 1989 (SI 1989/1156), as amended by SI 1990/1629 and SI 1992/339 and Environmental Protection (Prescribed Processes and Substances) Regulation 1991 (SI 1991/472)

- Urban Waste Water Treatment (England and Wales) Regulations 1994 (SI 1994/2841)

- The Waste Management Licensing Regulations 1994 (SI 1994/1056), The Waste Management Licensing (Amendment etc.) Regulations 1995 (SI 1995/288), The Waste Management Licensing (Amendment No. 2) Regulations 1995 (SI 1995/1995)

- Salmon and Freshwater Fisheries Act 1975

- Public Health Act 1936

Contaminated land

- Environment Act 1995 (EA 95) – Part II

Hazardous substances

- Environmental Protection Act 1990 – Part VIII

- Control of Pollution (Supply and Use of Injurious Substance) Regulations 1986 (SI 1986/902)

- EC Directive on Disposal of PCBs (96/59/EC)

- Proposed UK Action Plan for Phasing Out and Destructing PCBs

- Environmental Protection (Control of Injurious Substances) Regulation 1992 (SI 1992/31)

- Environmental Protection (Control of Injurious Substances) – No. 2 Regulation 1992 (SI 1992/1582)

- Environmental Protection (Control of Injurious Substances) Regulation 1993 (SI 1993/1)

- Environmental Protection (Control of Injurious Substances) – No. 2 Regulation 1993 (SI 1993/1643)

- Health and Safety at Work etc. Act 1974

- Notification of Installations Handling Hazardous Substances Regulation 1982 – NIHHS (SI 1982/1357)

- Dangerous Substances (Notification and Marking of Sites) Regulations 1990 (SI 1990/304)

- The Carriage of Dangerous Goods by Road Regulations 1996/2095

- The Carriage of Dangerous Goods by Rail Regulations 1996/2089

- The Carriage of Dangerous Goods by Road (Driver Training) Regulations 1996/2094

- Asbestos (Licensing) Regulations 1983 (SI 1983/1649)

- EU Regulation on the Evaluation and Control of Risks of Existing Substances (EEC 793/93)

- Notification of New Substances Regulations 1993 – NONS (SI 1993/3050)

- Chemical (Hazard Information and Packaging) Regulations 1994 – CHIP 2 (SI 1994/3247) and Chemical (Hazardous Information and Packaging for Supply) Amendment Regulations 1996

- Asbestos (Prohibitions) Regulations 1992 (SI 1992/3067)

- Control of Asbestos at Work Regulations 1987 (SI 1987/2115)

- Control of Pesticide Regulations 1986 (SI 1986/1510)

- EU Regulation EEC 594/91 on Substances that Deplete the Ozone Layer as amended by EEC 3952/92

- Environmental Protection (Non-Refillable Refrigerant Containers) Regulations 1994 (SI 1994/199)

- EU Regulation EEC 3093/94 on Substance that Deplete the Ozone Layer

- Radioactive Substances Act 1993 (RSA 93)

- Radioactive Substances (Appeals) Regulation 1990

- Radioactive Material (Road Transport – Great Britain) Regulations 1996

- The Packaging, Labelling and Carriage of Radioactive Material by Rail Regulation 1996

- Trans-frontier Shipment of Radioactive Waste Regulations 1993 (SI 1993/3031)

Incidents

- Control of Industrial Major Accident Hazards Regulations 1984 – CIMAH (SI 1984/1902)

- Public Information for Radiation Emergencies Regulation 1992 (SI 1992/2997)

- Notification of Cooling Towers and Evaporative Condensers Regulation 1992 (SI 1992/2225)

Statutory nuisance

- Environmental Protection Act 1990 (EPA 90) – Part III

- Statutory Nuisance (Appeals) Regulations 1995 (SI 1995/2644)

- Public Health Act 1936

- Clean Air Act (CAA 93)

- Noise and Statutory Nuisance Act 1993)

Planning

- Town and Country Planning Act 1990

- Town and Country Planning (Assessment of Environmental Effects) Regulations 1988 (SI 1988/119)

- Town and Country Planning (General Permitted Development) Order 1995 (SI 1995/418)

- Town and Country Planning (Environmental Assessment and Permitted Development) Regulations 1995 (SI 1995/417)

- Town and Country Planning (Environmental Assessment and Unauthorized Development) Regulations 1995 (SI 1995/2258)

- Planning (Hazardous Substances) Act 1990

- Planning and Hazardous Substances Regulations 1992 (SI 1992/656)

- Conservation (Natural Habitats etc.) Regulations 1994 (SI 1994/2716)

Litter

- Environmental Protection Act 1990 – Part IV

- Litter Controls Areas Order 1991

- Street Litter Controls Notices Order 1991

- Statutory Code of Practice on Litter and Refuse

Integrated pollution control

- Environmental Protection Act 1990 (EPA 90) – Part I

- Environmental Protection (Prescribed Processes and Substances) Regulation 1991 (SI 1991/472)

- Environmental Protection (Applications, Appeals and Registers) Regulation 1991 (SI 1991/507

Consents that are applicable to Dailuaine cover

- Cooling water discharge (x 2)

- Sludge disposal from treatment plant

- Disposal of heavy metal in sludge

- Treatment plant discharge

Annex 2

Check-lists for the initial environmental review

A.2.1 Check-lists for overview and general information

Description for Overview and General Information		
Check-list: The company or organization itself **Company name:** **Date prepared:** **Reviewed by:** **Page** **of**		
Questions to Ask	**Y/N**	**Comments**
1. What is the name of the company being reviewed?		
2. Are there parent or subsidiary companies associated with this organization?		
3. What is the ownership structure?		
4. How old is the company and how long has it been operating at its present location?		
5. What is the company culture (i.e. proactive, reactive, leaders, followers etc.)?		
6. Are they industry/trade association members? If so, which associations?		
7. What are the main activities, products or services of the company?		
Documents Collected		

Description for Overview and General Information		
Check-list: The site **Company name:** **Date prepared:** **Reviewed by:** **Page** **of**		
Questions to Ask	**Y/N**	**Comments**
1. What is the physical location of the site being reviewed (i.e. address, town, country etc.)? 2. How big is the site being reviewed (1 hectare, 5 square kilometres, 8 square miles)? 3. How many employees are there at the site? 4. Is the site located in one area or is it spread out? 5. What is the physical appearance of the site (i.e. is it well kept, clean and organized or is it unsightly and messy)?		
Documents Collected		

Description for Overview and General Information		
Check-list: Location of the site in relation to risk receptors and surroundings **Company name:** **Date prepared:** **Reviewed by:** **Page** **of**		
Questions to Ask	**Y/N**	**Comments**
1. Are there areas of natural significance in the vicinity (national parks, sensitive ecosystems, spawning/breeding grounds, wetlands, endangered species etc.)?		
2. Are there areas of cultural/historical significance in the vicinity (burial grounds, archaeological sites etc.)?		
3. Is there residential housing nearby?		
4. Are there any schools, hospitals, public parks, arenas or public attractions nearby?		
5. Is there a prevailing wind in the area?		
6. What is the land use to the north of the site?		
7. What is the land use to the east of the site?		
8. What is the land use to the south of the site?		
9. What is the land use to the west of the site?		
Documents Collected		

Description for Overview and General Information		
Check-list: Topography, hydrology and geography of the site **Company name:** **Date prepared:** **Reviewed by:** **Page of**		
Questions to Ask	**Y/N**	**Comments**
1. Could the site's actual or potential environmental aspects be exaggerated or mitigated by physical surroundings?		
2. Is the site in a valley, on a flood plain, on a hillside etc.?		
3. Is the site near a river, stream, sea, lake, estuary etc.?		
4. Have any previous hydrological or geological studies been undertaken at the site?		
5. Are there wells, aquifers or springs nearby?		
6. Is the area susceptible to seismic activity (earthquakes, tremors etc.)?		
7. If there are spills, leaks or uncontrolled discharges, where are they going to go?		
8. Is there potential for contaminating water sources?		
Documents Collected		

Description for Overview and General Information		

Check-list: Other local industry
Company name: **Date prepared:**
Reviewed by: **Page   of**

Questions to Ask	Y/N	Comments
1. Are there other industries or companies located nearby?		
2. Are they located upwind or downwind of your site?		
3. Are they located upstream or downstream of your site?		
4. Are they using the same watercourses as your site (up- and/or downstream)?		
5. Does your company co-operate in any manner with these other industries or companies?		
6. Has your company ever received complaints from, or complained to, these other industries or companies?		
7. Does your company know what these other industries or companies produce, emit and use in their operations?		
8. Do these other industries or companies know what is produced, emitted or used by your site?		
9. Do your site's environmental aspects and impacts have an obvious effect on any other neighbouring organizations?		
10. Do the environmental aspects or impacts of neighbouring organizations affect your site?		

Description for Overview and General Information		
Check-list: Other local industry		
Company name: **Date prepared:**		
Reviewed by: **Page** **of**		
Questions to Ask	**Y/N**	**Comments**
11. Is there the possibility that your site's environmental aspects and impacts are mitigated or exaggerated by the environmental aspects and impacts of a neighbouring site?		
Documents Collected		

Description for Overview and General Information		
Check-list: Site history **Company name:** **Date prepared:** **Reviewed by:** **Page** **of**		
Questions to Ask	**Y/N**	**Comments**
1. What activities, if any, have preceded present activities on the site? 2. Is it possible that a previous owner or occupier has contaminated the site? 3. Is there the possibility that your company's environmental aspects and impacts are mitigated or exaggerated by past activities of the site?		
Documents Collected		

A.2.2 Check-lists for review of environmental management practices

Review of Environmental Management Practices		
Check-list: Overall management **Company name:** **Reviewed by:**	**Date prepared:** **Page of**	
Questions to Ask	**Y/N**	**Comments**
1. What are the main management activities at the site?		
2. Under what management framework does the site operate?		
3. What is the company culture?		
4. Are there existing environmental management systems, procedures or policies?		
5. Are there internal/external pressures to improve the company's environmental performance?		
6. Is the development of an EMS seen as an important aspect of future business operations? (If so, why?)		
7. Is there a quality management system (QMS) at the site?		
8. Is there a health & safety system at the site?		
9. How well are your current management systems functioning?		
10. Do you anticipate integrating the EMS with existing management systems?		
11. Is the site currently engaged in any environmental projects or initiatives?		

Review of
Environmental Management Practices

Check-list: Overall management

Company name: **Date prepared:**

Reviewed by: **Page**   **of**

Questions to Ask	Y/N	Comments
12. Are you experiencing any recurring problems with your current management activities?		
Documents To Look For	**Y/N**	
▭ Quality management system documents		
▭ Worker health and safety documents		
▭ Existing EMS documents		
▭ Communication plans or flowcharts		
▭ List of on-going site projects		
▭ Any product LCA reports		
▭ Any product 'design for the environment' parameters or policies		

*For more comprehensive
check-list, see page 112*

Review of Environmental Management Practices		
Check-list: Environmental policy		
Company name:	**Date prepared:**	
Reviewed by:	**Page** **of**	
Questions to Ask	**Y/N**	**Comments**
1. Are there any statements of company commitment or mission statements?		
2. What are the company's goals, visions and values as told to employees and the public?		
3. Are there recognized codes of practice for the industry?		
4. What formal or informal policies are followed on site?		
5. Is there an existing formal or informal environmental policy?		
6. If there is an environmental policy: * Is it displayed for all to see? * Are employees aware of it? * Do people understand it? * Is it appropriate? * Is it realistic? * Is it followed?		
Documents To Look For	**Y/N**	
📖 Existing environmental policy		
📖 Existing quality policy		
📖 Existing occupational health & safety policy		
📖 Existing personnel satisfaction/ recognition policy		
📖 Purchasing policy		
📖 Subcontracting policy		
📖 Any other formally recognized site policies		

*For more comprehensive
check-list, see pages 104-105*

Review of Environmental Management Practices		
Check-list: Register of environmental aspects and impacts		
Company name:	**Date prepared:**	
Reviewed by:	**Page of**	
Questions to Ask	**Y/N**	**Comments**
1. Is there knowledge of the more significant environmental aspects of the site?		
2. How are the significant environmental aspects controlled?		
3. Have any reviews or assessments of impacts on the environment been conducted? (If so, what?)		
4. Does the company keep a register of environmental aspects and impacts relevant to the site?		
5. Have there been any significant accidents or incidents on site within recent years?		
6. Have there been any complaints regarding site operations within the last five years?		
Documents To Look For	**Y/N**	
📖 Register of environmental aspects and impacts (an analysis of environmental impacts of the production process)		
📖 A description of the organization's on-site processes		
📖 List of complaints made (over 5 yrs.)		
📖 Reports of any fines or breaches of compliance		
📖 Report of significant accidents or incidents on site (over last 5 yrs.)		

*For more comprehensive
check-list, see pages 128-129*

Review of
Environmental Management Practices

Check-list: Register of environmental legislation and regulations
Company name: **Date prepared:**
Reviewed by: **Page   of**

Questions to Ask	Y/N	Comments
1. Is the company aware of the legislation, regulations, authorizations, licenses or permits that it must comply with?		
2. Does the company keep a register of legislation and regulations relevant to the site?		
3. Are there any codes of practice that the company subscribes to?		
4. Does the company operate under permits or permit restrictions?		
5. Does the company operate under any authorizations, licenses or resulting restrictions?		
6. Has the company been fined or warned for past regulatory/legislative noncompliance?		
7. Is the company aware of any present/ past regulatory/legislative noncompliance?		
8. Does the company follow a procedure to identify and have access to environmental legislation and regulations that affects it?		
9. Are there regulations specific to the company's activities?		
10. Are there regulations specific to the company's products or services?		

Review of
Environmental Management Practices

Check-list:	Register of environmental legislation and regulations
Company name:	**Date prepared:**
Reviewed by:	**Page** **of**

Questions to Ask	**Y/N**	**Comments**
11. Are there regulations specific to the company's industry?		
12. Is the company aware of major international conventions relevant to it activities (i.e. Montreal Protocol, Basel Convention and Agenda 21)?		
13. Have internal staff performed any legislative or regulatory inspections? (If so, by whom and when?)		
14. Have local authorities performed any site inspections?		
15. Have independent companies or consultants performed any site inspections? (If so, by whom and when?)		
Documents To Look For	**Y/N**	
📖 Register of relevant legislation and regulations		
📖 Industry, company or company group codes of practice		
📖 List of permits required for operation		
📖 License requirements		
📖 List of relevant authorities applicable to the company		
📖 List of industry or professional associations applicable to the company		
📖 An overview of the applicable regulations and agreements with authorities		
📖 Documentation of emissions		
📖 Details of any breaches of legislation and regulations		

*For more comprehensive
check-list, see pages 138-139*

Review of **Environmental Management Practices**		
Check-list: Objectives and targets		
Company name: **Date prepared:**		
Reviewed by: **Page** **of**		
Questions to Ask	**Y/N**	**Comments**

1. Does the site have any environmental objectives or targets associated with their operations?
 	
2. If so:
 	
    *   What are they?
 	
    *   Are the objectives directly related to significant environmental impacts or stated policy goals?
 	
    * Are the set targets being achieved?		

Documents To Look For	**Y/N**	
📖 List of environmental objectives and targets set for the site 📖 List of overall site objectives and targets		

*For more comprehensive
check-list, see page 148*

Review of Environmental Management Practices		
Check-list: Environmental management programmes **Company name:** **Date prepared:** **Reviewed by:** **Page** **of**		
Questions to Ask	**Y/N**	**Comments**
1. Are there programmes in place to assure that stated objectives and targets are met? 2. If so, what do they involve? 3. What notable environmental investments have been made over the last five years?		
Documents To Look For	**Y/N**	
📖 Any documented plans or programmes for meeting stated objectives and targets 📖 Information of on-going environmental projects 📖 List of notable 'environmental investments' over the last 6 years		

*For more comprehensive
check-list, see page 156*

Review of Environmental Management Practices		
Check-list: Structure and responsibility **Company name:** Date prepared: **Reviewed by:** Page of		
Questions to Ask	**Y/N**	**Comments**
1. What are the lines of responsibility on site? 2. Are there environmentally-related job descriptions and responsibilities? (If so, what are they?) 3. Are there documented procedures for all activities, products and processes that have, or could have if uncontrolled, a significant direct or indirect impact on the environment?		
Documents To Look For	**Y/N**	
📖 Site organization chart 📖 Job descriptions 📖 Organizational chart showing environmental responsibilities		

*For more comprehensive
check-list, see pages 173-174*

Review of
Environmental Management Practices

Check-list: Training, awareness and competence
Company name: **Date prepared:**
Reviewed by: **Page   of**

Questions to Ask	Y/N	Comments
1. Are there existing environmental training programmes? (If so, describe.)		
2. Has there been any previous environmental training? (If so, describe)?		
3. Are there other on-going or regular training programmes (not environmentally-related)?		
4. How is training conducted at the site?		
5. Does the company show videos, case study examples or practical demonstrations for training purposes?		
6. Has the company engaged in WH & S or QMS training?		
7. What is the overall level of education and training in the company?		
Documents To Look For	**Y/N**	
📖 List of training conducted on site		

*For more comprehensive
check-list, see pages 180-181*

Review of Environmental Management Practices

Check-list: Environmental communication

Company name: **Date prepared:**

Reviewed by: **Page**   **of**

Questions to Ask	Y/N	Comments
1. Are there existing lines of communication or communication procedures in the company?		
2. Who are the company's main stakeholders?		
3. How does the site communicate externally?		
4. How does the site communicate internally?		
5. What is the site's relationship with the local community?		
6. Have there been any recent complaints from the local community? (If so, why?)		
7. What is the company's relationship with the local authority?		
8. Does the site make a point of using suppliers with certified QMSs or EMSs?		
9. Has the site prepared an environmental report or statement? (If so, is it for internal or external distribution?)		
Documents To Look For	**Y/N**	
📖 Record of any complaints received and actions taken in response to complaints		
📖 Supplier questionnaires		
📖 Site environmental report or statement		
📖 Environmental reports		

*For more comprehensive
check-list, see pages 191-192*

Review of Environmental Management Practices		
Check-list: EMS documentation **Company name:** **Date prepared:** **Reviewed by:** **Page** **of**		
Questions to Ask	**Y/N**	**Comments**
1. What environmentally-related documentation is generated or used on site?		
Documents To Look For	**Y/N**	
📖 Environmental management system documentation		
📖 Site procedures and work instructions		
📖 List of documents kept on site		
📖 Cross-referenced list linking the documentation to the related requirements of the standard		

*For more comprehensive
check-list, see page 197*

Review of
Environmental Management Practices

Check-list: Document control

Company name: **Date prepared:**

Reviewed by: **Page** **of**

Questions to Ask	Y/N	Comments
1. Is there a documented procedure for controlling documents necessary for the development, implementation and maintenance of your management system?		
2. Is there a documented procedure for preparation, storing, issuing, amending and revising of management documentation?		
3. Are the documents on site kept up to date?		
4. Are the necessary management documents legible and easily located by those who require them?		
Documents To Look For	**Y/N**	
📖 Procedure or methodology for controlling documents		

*For more comprehensive
check-list, see page 203*

Review of
Environmental Management Practices

Check-list: Operational control

Company name: **Date prepared:**

Reviewed by: **Page**   **of**

Questions to Ask	Y/N	Comments
1. Are the activities, products and processes associated with significant environmental impacts effectively controlled?		
2. What methods are used to achieve operational control of activities, products and processes that could lead to a significant environmental impact?		
3. Does the site have documented work instructions and procedures for all activities, products and processes that could lead to a significant environmental impact?		
4. Is there a process of approval for all planned activities, products, processes and procurements?		
Documents To Look For	**Y/N**	
📖 List of site procedures and work instructions		

*For more comprehensive
check-list, see page 209*

Review of Environmental Management Practices		
Check-list: Emergency preparedness and response **Company name:** Date prepared: **Reviewed by:** Page of		
Questions to Ask	**Y/N**	**Comments**
1. Is the potential for accidents and/or emergency situations associated with the site's activities, products and process identified and controlled? 2. Are there procedures in place to appropriately 'respond to' and minimize the environmental impact of accident and emergency situations? (If so, are they regularly tested and revised if necessary?)		
Documents To Look For	**Y/N**	
📖 List of procedures dealing with emergency situations		

*For more comprehensive
check-list, see page 215*

Review of
Environmental Management Practices

Check-list: Monitoring and measuring

Company name: **Date prepared:**

Reviewed by: **Page   of**

Questions to Ask	Y/N	Comments
1. Is monitoring and measuring practised on a regular basis?		
2. Are there documented procedures for monitoring and measuring?		
3. What is the frequency of monitoring and measuring?		
4. Are monitoring and measurement records kept? If so where?		
5. Who is responsible for monitoring and measuring?		
6. Are all operations with potential significant environmental impact monitored and measured?		
7. Are there monitoring and measuring programmes for emissions to air?		
8. Are there monitoring and measuring programmes for discharges to land?		
9. Are there monitoring and measuring programmes for effluents to water?		
10. Are there monitoring and measuring programmes for noise or vibration?		
11. What monitoring and measuring equipment is used?		
12. How and when is the equipment calibrated and tested and is this procedure documented?		

Review of
Environmental Management Practices

Check-list: Monitoring and measuring

Company name: **Date prepared:**

Reviewed by: **Page   of**

Questions to Ask	Y/N	Comments
13. Are laboratory services used (internal or external)?		
14. What analytical standards are used?		
15. Have there been internal complaints about the company's environmental performance?		
16. Is there a register of internal complaints?		
Documents To Look For	**Y/N**	
▭ Monitoring and measuring records made and kept		
▭ List of monitoring and measuring equipment used		
▭ Maps indicating where monitoring and measuring takes place		
▭ Any equipment calibration and testing records		
▭ Data acceptance criteria		
▭ Register of internal complaints		

*For more comprehensive
check-list, see page 223*

Review of
Environmental Management Practices

Check-list: Nonconformance, correction and prevention
Company name: **Date prepared:**
Reviewed by: **Page** **of**

Questions to Ask	Y/N	Comments
1. Does the site conform to its stated environmental policy, objectives and targets, and other environmental requirements to which it subscribes?		
2. Is there a procedure in place to correct and prevent the future occurrence of the nonconformance?		
3. Is there a record of nonconformances?		
Documents To Look For	**Y/N**	
📖 Procedure to handle situations of nonconformance		
📖 Records of past nonconformances		

*For more comprehensive
check-list, see page 229*

Review of Environmental Management Practices		
Check-list: Environmental records **Company name:** Date prepared: **Reviewed by:** Page of		
Questions to Ask	**Y/N**	**Comments**
1. What environmental records (if any) are kept on site? 2. Is there a written account of correspondence with licensing authorities, waste removal companies and delivery of goods?		
Documents To Look For	**Y/N**	
📖 List of records kept on site 📖 Record of correspondence with licensing authorities 📖 Waste removal records 📖 Hazardous waste removal records		

*For more comprehensive
check-list, see pages 256-259*

Review of Environmental Management Practices		
Check-list: EMS audit **Company name:** Date prepared: **Reviewed by:** Page of		
Questions to Ask	**Y/N**	**Comments**
1. Is there an existing environmental auditing programme? 2. Have there been previous audits performed? 3. If so: * How where these performed? * By whom were they performed? * When was(were) the audit(s) performed? * What was the scope of the audits(s)? * What is the frequency of the audit(s)? * Are the audit findings and recommendations of the audit(s) acted on?		
Documents To Look For	**Y/N**	
📖 Previous environmental or EMS audits 📖 Reports generated by the authorities assessing the site's efficacy in controlling its environmental impact 📖 Internal site assessments of existing management systems		

*For more comprehensive
check-list, see page 265*

Review of Environmental Management Practices		
Check-list: Management review **Company name:** **Date prepared:** **Reviewed by:** **Page** **of**		
Questions to Ask	**Y/N**	**Comments**

1. Is there a management review process of assessing whether the site's environmental management activities maintain their:
 	
  * Adequacy?
 	
  * Suitability?
 	
  * Effectiveness in relation to the organization's overall intentions with respect to improved environmental performance? (If so, is it documented?)		

Documents To Look For	**Y/N**	
📖 Management reports outlining changes to current practice or introducing new initiatives 📖 Internal memos confirming changes to management system practices		

A.2.3 Check-lists for the review of activities, products and processes

<table>
<tr>
<td colspan="4" align="center">

**Description of the Main Processes
from an Environmental Point of View**

</td>
</tr>
<tr>
<td colspan="4">

Main process reviewed: **Check-list:**   Water use
Company name: **Date prepared:**
Reviewed by: **Page**   **of**

</td>
</tr>
<tr>
<td align="center">**Questions to Ask**</td>
<td align="center">**Y/N**</td>
<td colspan="2" align="center">**Comments**</td>
</tr>
<tr>
<td>

1. Is there water used in the process step? If so what is it used for?

2. What quantity of water is used?

3. What is the source and cost of the water used?

4. Are there any water use permits, consents or authorizations in the process step?

5. If so, what are those permits, consents or authorizations?

6. Is water used for indirect activities such as cleaning?

7. Are water use minimization or cleaner production techniques practised in the process step? If so, what are they?

Hint: *If water is used in the process, keep an eye out for running taps, leaks etc.*

</td>
<td></td>
<td colspan="2"></td>
</tr>
<tr>
<td align="center">**Documents To Look For**</td>
<td align="center">**Y/N**</td>
<td colspan="2" align="center">**Comments**</td>
</tr>
<tr>
<td>

📕   Water bills

📕   Water use records

📕   Water use permits, consents or authorizations

</td>
<td></td>
<td colspan="2"></td>
</tr>
</table>

Description of the Main Processes
from an Environmental Point of View

Main process reviewed: **Check-list:**   Energy use

Company name: **Date prepared:**

Reviewed by: **Page**   **of**

Questions to Ask	Y/N	Comments
1. Is energy used in the process step?		
2. What quantity of energy is used?		
3. What is it used for and what is the source and cost of the energy used?		
4. Are there pumps, compressors, motors, boilers used in the process step? If so, which and what quantity?		
5. Is an energy saving or minimization programme practised in the process step and if so, what?		
Hint: *If pumps, compressors or motors are used in the process, try to assess their efficiency as tremendous efficiency improvements can usually be made in this area!*		
Documents To Look For	**Y/N**	**Comments**
📖 Energy bills		
📖 Energy use records		

Description of the Main Processes		
from an Environmental Point of View		

Main process reviewed:	**Check-list:** Chemical use
Company name:	**Date prepared:**
Reviewed by:	**Page** **of**

Questions to Ask	**Y/N**	**Comments**
1. Are chemicals used in the process step, and if so what are they?		
2. What quantities of chemicals are used?		
3. What is the source and cost of the chemicals used?		
4. Do the chemicals used have any noteworthy environmental, health or safety implications?		
5. Is material safety data information kept for the chemicals used?		
6. Are the chemicals properly stored?		
7. Are any of the chemicals used in the process step legislated or regulated?		
8. Do any of the chemicals used in the process step require permits, consents or authorization? If so, which ones?		
9. Are chemical use minimization or cleaner production techniques practised in the process step? If so, what are they?		
Hint: *Pay special attention to chemical storage. Look in storerooms, warehouses, behind doors etc. to see if chemicals are stored, how they are stored, if they are properly labelled, and so on!*		
Documents To Look For	**Y/N**	**Comments**
▢ List of chemicals used, quantities, cost and suppliers		
▢ Chemical safety data sheets		

Description of the Main Processes from an Environmental Point of View			

Main process reviewed: **Check-list:**   Raw material use
Company name: **Date prepared:**
Reviewed by: **Page**   **of**

Questions to Ask	Y/N	Comments
1. What are the main raw materials used in the process step?		
2. What quantities are used?		
3. What are sources and costs of the raw materials used?		
4. Are minimization or cleaner production techniques practised in the process step, if so, what are they?		
Documents To Look For	**Y/N**	**Comments**
📖 List of raw materials used, their quantities, costs and suppliers		
📖 Raw material safety data information		
📖 Raw material purchase order forms		
📖 Site raw material inventory lists		

Description of the Main Processes from an Environmental Point of View		

Main process reviewed: **Check-list:**   Storage

Company name: **Date prepared:**

Reviewed by: **Page**   **of**

Questions to Ask	Y/N	Comments
1. What raw materials, inputs, outputs, finished or partly finished products are stored in the process step?		
2. Where are raw materials, inputs, outputs, finished or partly finished products stored in the process step?		
3. Is an inventory list of the items stored in the process step maintained? If so, where and how often is it updated?		
4. Do any of the items stored in the process require permits, consents or authorizations? If so, what items are they and what are the permits, consents or authorizations required?		
5. Are any of the items stored in the process step legislated or regulated? If so, what items are they and what is the legislation or regulation?		
6. Are any of the items stored in the process step monitored? If so, which ones and how are they monitored?		
7. Is any of the monitoring of items stored in the process step mandatory? If so, which items?		

Description of the Main Processes from an Environmental Point of View		
Main process reviewed: **Company name:** **Reviewed by:**	**Check-list:** Storage **Date prepared:** **Page of**	
Questions to Ask	**Y/N**	**Comments**
8. Are there oil, gas, diesel or other hazardous substances stored in the process? If so, what is stored and where is it stored? **Hint:** *Don't forget to assess whether substances indirectly associated with a production process – such as oil, gasoline, diesel, lubricants, cleaning solutions etc. are stored somewhere on site!*		
Documents To Look For	**Y/N**	**Comments**
📖 Storage inventory lists 📖 Storage location maps 📖 Storage permits, consents, authorizations 📖 Storage noncompliance records		

Description of the Main Processes from an Environmental Point of View		
Main process reviewed: **Company name:** **Reviewed by:**	**Check-list:** Effluents to water **Date prepared:** **Page of**	
Questions to Ask	**Y/N**	**Comments**
1. Is effluent discharged in the process step? If so, what is the discharge?		
2. Where is effluent discharged to and from where does the effluent originate and in what quantities?		
3. Is effluent treated in the process step? If so, what effluent, how is it treated, where and by whom is it treated?		
4. Is there a treatment facility in the process step? If so, how old is the facility and what are the procedures in case of facility failures?		
5. Do any of the discharges from this process step require permits, consents or authorizations? If so, what are the discharges and what is the permit, consent or authorization required?		
6. Are any of the discharges in the process step legislated or regulated? If so, which ones?		
7. Are any of the discharges in the process step monitored? If so, which ones, when and how are they monitored?		

Description of the Main Processes
from an Environmental Point of View

Main process reviewed:	**Check-list:** Effluents to water
Company name:	**Date prepared:**
Reviewed by:	**Page** **of**

Questions to Ask	Y/N	Comments
8. Is any discharge monitoring in the process step mandatory? If so, which discharge and from what source?		
9. Are any of the process discharges recorded? If so, where, when and how are they recorded?		
10. What is the age, nature of the drainage systems in the process step?		
11. Are there effluent retention, storage or sedimentation tanks (etc.) used in the process step? If so, where are they?		
12. Are there weirs, separators, septic tanks, etc. in the process step? If so, where, what are they used for and what occurs if they fail, overflow or malfunction?		
13. What is the cost of effluent discharges from the process step?		
14. Is there the possibility of accidental spills, leaks or uncontrolled discharges in this process step? If so, where, what could be spilled or leaked and what is the receiving water course?		
15. Is there any discharge control/abatement equipment used in the process step? If so, what is used and where is it used?		

Description of the Main Processes from an Environmental Point of View		
Main process reviewed: **Check-list:** Effluents to water **Company name:** **Date prepared:** **Reviewed by:** **Page** **of**		
Questions to Ask	**Y/N**	**Comments**
16. Are effluent discharge minimization or cleaner production techniques practised in the process step? If so, what are they? **Hint:** *Pay special attention to floor drains and outside drains and drains near things like chemical stores. Note where drains are and everything that could possibly go down them, both intentionally and accidentally!*		
Documents To Look For	**Y/N**	**Comments**
📖 Discharge records 📖 Discharge permits, consents or authorizations 📖 Discharge monitoring and/or analysis records 📖 Discharge noncompliance records 📖 Site/process drainage diagrams		

Description of the Main Processes
from an Environmental Point of View

Main process reviewed:	**Check-list:** Emissions to air
Company name:	**Date prepared:**
Reviewed by:	**Page** **of**

Questions to Ask	Y/N	Comments
1. Are there emissions to air from the process step? If so, what are they?		
2. What is the quantity of emissions?		
3. Are air emissions controlled or treated in the process step? If so, how, by whom, where and at what cost?		
4. Are any of the emissions from the process step legislated or regulated?		
5. Do any of the emissions from the process step require permits, consents or authorizations? If so which ones?		
6. Are any of the emissions from the process step monitored? If so, where, when and how are they monitored?		
7. Is any emission monitoring in the process step mandatory? If so, what is monitored and how is it monitored?		
8. Are any of the emissions from the process step recorded? If so, which emissions are they, when and how are they monitored?		
9. Are there any noticeable odours in the process step? If so, what are the odours and where do they originate?		

Description of the Main Processes from an Environmental Point of View		
Main process reviewed: **Check-list:** Emissions to air **Company name:** **Date prepared:** **Reviewed by:** **Page** **of**		
Questions to Ask	**Y/N**	**Comments**
10. Is there any emission control/abatement equipment used in the process step? If so, what technique is it and where is it used? 11. Are emissions minimization or cleaner production techniques practised in the process step. If so, what are they?		
Documents To Look For	**Y/N**	**Comments**
📖 Emission records 📖 Emission permits, consents or authorizations 📖 Ventilation diagrams 📖 Emission monitoring and/or analysis records 📖 Emission noncompliance records		

Description of the Main Processes
from an Environmental Point of View

Main process reviewed:	**Check-list:** Disposal to land
Company name:	**Date prepared:**
Reviewed by:	**Page** **of**

Questions to Ask	Y/N	Comments
1. Is there solid waste generated in this process? If so, what kind of waste is it, where is it disposed, in what quantities and at what cost?		
2. What is the final destination of solid waste from the process step and how is it transported to that destination?		
3. Is solid waste stored, treated, separated, recycled or reclaimed in the process step? If so, what waste and in what quantity?		
4. Are there off-specification product wastes in the process step. If so, what wastes, in what quantity and at what cost?		
5. Is packaging disposed of in the process step? If so, what sort of packaging, in what quantities and at what cost?		
6. Is any of the waste disposed in the process step legislated or regulated? If so, what waste and what is the legislation or regulation?		
7. Do any of the process wastes disposed require permits, consents or authorizations? If so, what wastes and what are the permits, consents or authorizations required?		

Description of the Main Processes from an Environmental Point of View			
Main process reviewed: **Company name:** **Reviewed by:**		**Check-list:** Disposal to land **Date prepared:** **Page** **of**	
Questions to Ask	**Y/N**	**Comments**	
8. Is the waste disposed in the process step monitored. If so, where, when, how is that waste monitored? 9. Is any of the waste disposal monitoring in the process step mandatory? If so, what? 10. Is there any waste control/abatement equipment used in the process step? If so what is it and where is it used? 11. Are waste minimization or cleaner production techniques practised in the process step? If so, what are they?			
Documents To Look For	**Y/N**	**Comments**	
📖 Waste disposal records 📖 Waste disposal permits, consents or authorizations 📖 Waste disposal monitoring and/or analysis records 📖 Disposal noncompliance records			

Description of the Main Processes from an Environmental Point of View		
Main process reviewed: **Company name:** **Reviewed by:**	**Check-list:** **Date prepared:** **Page** **of**	Hazardous, special, restricted products & waste
Questions to Ask	**Y/N**	**Comments**
1. Does the process step generate, use, store or dispose of any special or hazardous products or wastes? If so, what is that product or waste, what quantities are generated and what makes that product or waste special or hazardous?		
2. If special or hazardous products or wastes are stored in the process step, how are they stored and is a storage inventory list maintained?		
3. If there are special or hazardous products or wastes generated in the process step, how are they disposed of and by whom?		
4. Do any of the special or hazardous products or wastes generated in the process step require permits, consents or authorizations? If so what are the products or waste and what are the permits, consents or authorizations required?		
5. Are any of the special or hazardous products or wastes associated with the process step legislated or regulated? If so, what is that product or waste and what is the legislation or regulation?		
6. Are any of the special or hazardous products or wastes generated in the process step monitored? If so, what is that product or waste and how is it monitored?		

Description of the Main Processes from an Environmental Point of View			

**Description of the Main Processes
from an Environmental Point of View**

Main process reviewed:		Check-list:	Hazardous, special, restricted products & waste
Company name:		Date prepared:	
Reviewed by:		Page of	

Questions to Ask	Y/N	Comments
7. Is any of the monitoring of special or hazardous products or wastes associated with the process step mandatory?		
8. Is there the possibility of a spill or leakage of any of the special or hazardous material? If so, what is the material, where could it spill or leak from and where would it spill or leak to?		
9. Are special or hazardous material use and waste minimization or cleaner production techniques practised in the process step? If so, what are they?		
Hint: *The more likely a substance or activity is to have a significant environmental aspect or impact, the more likely it is that people will tell you what they think should be happening rather than what is happening. Thus with issues like chemicals, hazardous substances, or the risk of accident or emergency, you should always try to double-check the information you receive or ensure that what is being said is actually true!*		
Documents To Look For	**Y/N**	**Comments**
📖 Hazardous/special waste records		
📖 Hazardous/special waste permits, consents or authorizations		
📖 Hazardous/special waste monitoring and/or analysis records		
📖 Hazardous/special disposal noncompliance records		

Description of the Main Processes
from an Environmental Point of View

Main process reviewed:	**Check-list:** Other
Company name:	**Date prepared:**
Reviewed by:	**Page** of

Questions to Ask	**Y/N**	**Comments**
1. Is there significant noise pollution from this process step? If so, what is its source and magnitude? 2. Are there significant vibrations from this process? If so, what are their sources and magnitudes? 3. Is there any significant on-site transportation? If so what? 4. Is there any significant off-site transportation associated with the activities, products or processes being reviewed?		
Documents To Look For	**Y/N**	**Comments**

Annex 3

List of general environmental impact descriptions

Aspect reference numbers	General aspects	Impact description
WU	**Water use**	
WU01	Use of municipal water	Depletion of limited potable water resources. Water is one of the prime resources of life. Only 3% of world's water is fresh and only 0.003% of world's water is suitable for drinking, irrigation or industry.
WU02	Use of water from surrounding water courses	Impact can affect local environment and water ecology. Depletion of limited potable water resources. Water is one of the prime resources of life. Only 3% of world's water is fresh and only 0.003% of world's water is suitable for drinking, irrigation or industry.
WU03	Other water use	Depletion of limited potable water resources. Water is one of the prime resources of life. Only 3% of world's water is fresh and only 0.003% of world's water is suitable for drinking, irrigation or industry.
EU	**Energy use**	
EU01	Use of natural gas for energy	Depletion of non-renewable fossil fuel resources. Global gas reserves estimated to be 80 yrs. Combustion leads to emission of VOCs, NO_x, Methane and CO_2 and thus air pollution, acidification, greenhouse gases and global warming.
EU02	Use of oil for energy	Depletion of non-renewable fossil fuel resources. Global oil resources estimated to be 70-80 yrs. Combustion leads to emission of VOCs, NO_x, SO_x, CO_2 and thus air pollution, acidification, greenhouse gas production and global warming.

Aspect reference numbers	General aspects	Impact description
EU03	Use of coal for energy	Depletion of non-renewable fossil fuel resources. Global reserves estimated at 200 yrs. Coal is the dirtiest fossil fuel and produces VOCs, NO_x, SO_x, CO_2 and thus air pollution, acidification, greenhouse gas production and global warming.
EU04	Use of other fossil fuels	Depletion of non-renewable fossil fuel resources. Combustion leads to emission of VOCs, NO_x, SO_x, CO_2 and thus air pollution, acidification, greenhouse gas production and global warming.
EU05	Use of fuel for transportation	Transportation (fossil fuels) is a source of VOCs, NO_x, SO_2, CO_2 emissions and thus air pollution, acidification, greenhouse gases and global warming. Transportation is the source of 50% average air pollution, CO_2 (14%) and CFC (28%).
EU06	Use of energy from nuclear sources	Depletion of non-renewable natural resources. Impact through power plant cooling water and raw material use (uranium 235, plutonium 239) and disposal of radioactive waste.
EU07	Use of energy from hydro-electric sources	Hydropower is one of the world's cleanest sources of energy. It is renewable and devoid of CO_2 or other air emissions in the process. Hydro accounts for approximately 20% world energy. Larger hydro dams can have severe effect on the local ecology and surrounding areas from deforestation, loss of biodiversity, social impact.
EU08	Use of energy from wind sources	Wind power is one of the world's 'cleanest' forms of energy. Wind power generates no CO_2 or other air emissions, is renewable and does not require cooling waters. Land occupied by wind farms can be used for other purposes such as farming and grazing. Wind power has noise and visual impact.
EU09	Use of energy from solar sources	Solar electricity (photovoltaic cells) is one of the world's cleanest forms of renewable energy. Solar power produces no CO_2 during use and air and water pollution during use is extremely small. Solar cells can be unsightly and production can lead to chemical effluent discharges.
EU10	Use of energy from mixed sources	Impact will vary depending on mix. Impact can be determined by estimating impact using the impact descriptions for aspects EU01 to EU10 above.
EU11	Other energy use	Impact will vary depending on other source and local environment.

Aspect reference numbers	General aspects	Impact description
CU	**Chemicals use**	
CU01	Use of hazardous, restricted or special chemicals or materials	For specific toxicity, health risks, composition and environmental impact, refer to chemical databases at the handbook web site http://www.entropy-international.com/handbook/.
CU02	Use of acidic chemicals	Acids (pH less than 7) can impact on local ecology and affect human health. In solution, i.e. water, acids split into H+ ions and can be highly corrosive. In solutions with metals, acids form hydrogen gas and react with bases to form salts.
CU03	Use of basic chemicals	Alkalis or bases (pH more than 7) can impact on local ecology and affect human health. As solution in water, creates hydrogen oxide (HO-).
CU04	Use of solvents	Liquid solvents can contaminate ground and groundwater resources (1 litre can contaminate 100 million litres of drinking water). Some are VOCs and contribute to low-level air pollution (affect human health and vegetation) global warming and ozone depletion.
CU05	Use of hydraulic oils, lubricants, greases etc.	Depletion of natural resources. Disposal can contaminate local environment.
CU06	Other chemical use	Impact depends on chemical used. For specific toxicity, health risks, composition and environmental impact, refer to chemical databases at the handbook web site http://www.entropy-international.com/handbook/.
RU	**Raw material ue**	
RU01	Use of raw materials (hazardous, special or restricted and on 'red list')	Impact depends on substance used. For specific toxicity, health risks, composition and environmental impact, refer to chemical databases at the handbook web site http://www.entropy-international.com/handbook/.
RU02	Use of raw materials (not on 'red list')	Depletion of natural resources, contribution to solid waste.
RU03	Use of packaging materials	Depletion of natural resources, contribution to solid waste.
RU04	Use of office materials (paper, toner etc.)	Use of raw materials, contribution to solid waste if not recycled.
RU05	Use of construction materials	Depletion of natural resources, contribution to solid waste.
RU06	Other raw material use	Depletion of natural resources, contribution to solid waste.

Aspect reference numbers	General aspects	Impact description
ST	**Storage on site**	
ST01	Storage of chemicals	Potential for accidental, unintentional or undetected spills, leaks or discharge of stored chemicals and could interact with other stored materials, affect human health and impact on local or global environment.
ST02	Storage of raw material	Potential for accidental, unintentional or undetected discharges or could interact with other stored materials, affect human health and impact on local or global environment.
ST03	Storage of 'red list' materials	Risk of impact from spill, leak etc. Impact depends on substance stored. For specific toxicity, health risks, composition and environmental impact, refer to chemical databases at the handbook web site http://www.entropy-international.com/handbook/.
ST04	Storage of waste on site	Risk of impact from spill, leak etc. Impact depends on waste composition.
ST05	Storage of special waste on site	Potential for spills, leaks or discharge and could interact with other materials, affect human health and impact on environment. Impact exaggerated as special waste is often flammable and may have toxic, mutagenic, carcinogenic or other irreversible effects.
ST06	Other storage	Risk of impact from spill, leak etc. Impact depends on material stored.
EW	**Effluents to water**	
EW01	Discharge to treatment facility	Treatment usually includes reduction of organic or suspended matter in wastewater. Organic matter is typically removed through oxidization and aerobic digestion. Produces sludge bi-product needing to be disposed of or recycled as fertilizer.
EW02	Controlled discharge of treated effluent to water courses	Impact will vary depending on both receiving waters and composition of effluent. Impact arises from alterations to natural ecosystem.
EW03	Controlled discharge of untreated effluent to water courses	Impact will vary depending on both receiving waters and composition of effluent. Impact arises from alterations to natural ecosystem.
EW04	Uncontrolled discharge of treated effluent to water courses	Impact will vary depending on both receiving waters and composition of effluent. Impact arises from alterations to natural ecosystem.

Aspect reference numbers	General aspects	Impact description
EW05	Uncontrolled discharge of untreated effluent to water courses	Impact will vary depending on both receiving waters and composition of effluent. Impact arises from alterations to natural ecosystem.
EW06	Discharge of 'red list' substances	Impact will vary depending on substance discharged. For full impact description see web site at http://www.entropy-international.com/handbook/. All such discharges are likely to affect growth rate of aquatic diatoms disturbing entire aquatic food chains and impacting on ecology of the receiving environment. Some compounds can be lethal to resident natural life and all future users of receiving waters.
EW07	Other discharges	Impact will vary depending on both receiving waters and composition of effluent. Impact arises from alterations to natural ecosystem.
EA	**Emissions to air**	
EA01	Emission of process gases/heat	Affects working environment, contributes to low-level air pollution and global warming.
EA02	Emission of flue gases	Contribution to the greenhouse effect, activity can change ecology of local environment.
EA03	Emission of NO_x	Conversion of nitrogen oxides (NO_x) to nitric acids contributes to acid rain, NO_x contributes to photochemical smog, contributes to global warming and stratospheric ozone depletion. Impact to health, local and global environment.
EA04	Emission of SO_x	Conversion of sulphur oxides (SO_x) to sulphuric acids contributes to acid rain. SO_x contributes to photochemical smog. Impact to health, local and global environment. SO_2 has a direct correlation to asthma, bronchitis and damage to plants and aquatic ecosystems.
EA05	Emission of CO and CO_2	Gases produced from combustion of organic matter. Estimated to be responsible for 50% of greenhouse gases and global warming. CO (from incomplete combustion) is poisonous if inhaled and responsible for more severe chemical poisonings than any other single agent. Cigarette smoking is the largest source of human exposure to CO.
EA06	Emission of particulate matter	Deposition of fine solid or liquid droplets to the atmosphere. Contribute to air pollution directly and indirectly as synergists or carriers of other pollutants. Can affect health and local environment.

Aspect reference numbers	General aspects	Impact description
EA07	Emission of dust or raw material	Contribute to air pollution directly and indirectly as synergists or carriers of other pollutants. Can affect health and local environment.
EA08	Emission of VOCs	Volatile Organic Compounds (propane, benzene, CFCs, solvents etc.) evaporate readily and contribute to low level air pollution, global warming and can cause ozone depletion. VOCs can affect human respiratory health, crops and vegetation.
EA09	Emission of 'red list' substances	Impact depends on substance emitted. For specific toxicity, health risks, composition and environmental impact, refer to chemical databases at the handbook web site http://www.entropy-international.com/handbook/.
EA10	Emissions from transport	Transportation (fossil fuels) is source of VOCs, NO_x, SO_2, CO_2 emissions and thus air pollution, acidification, greenhouse gases and global warming. Considerable source of world CO_2 (14%) and CFC (28%) emissions and source of 50% average air pollution.
EA11	Other emissions	Impact will vary according to emission and surrounding areas.
DL	**Disposal to land**	
DL01	Disposal to municipal landfill	Land use, degradation and contamination. Contribution to acidic landfill leachate high in Biological Oxygen Demand (BOD), ammonia, organic nitrogen, volatile fatty acids and other toxins. Landfill gases contribute to greenhouse effect and global warming. Affects local ecology and environment and is malodorous.
DL02	Disposal to site landfill	Land use, degradation and contamination. Contribution to acidic landfill leachate high in Biological Oxygen Demand (BOD), ammonia, organic nitrogen, volatile fatty acids and other toxins. Landfill gases contribute to greenhouse effect and global warming. Affects local ecology and environment and is malodorous.
DL03	Disposal to incineration	Emission of VOCs, NO_x, SO_x, CO_2 and particulate matter to atmosphere and thus contribute to ozone depletion, air pollution, greenhouse gas production and global warming.
DL04	Disposal to recycling, reclamation or re-use	Recycling, reclamation and re-use reduce impact from disposal but impact of production and use are still of noteworthy mention and should be considered.

Aspect reference numbers	General aspects	Impact description
DL05	Disposal of special or hazardous waste	Land use, degradation and contamination. Disposal of metallic compounds accelerates contamination and hazards to groundwater and local ecology.
DL06	Disposal of contaminated soil/land	Land contamination is one of the most severe global issues. Poisoning from landfills, heavy metals, toxins, chemicals, acids, alkalis etc. is linked to dozens if not hundreds of subsequent environmental and social health issues.
DL07	Other discharges	Impact will vary depending on discharge and local environment.
OT	**Other**	
OT01	Vibrations	Impact will vary depending on areas affected.
OT02	Noise, smell	Excessive or prolonged exposure to noise (typically more than 8 hrs above 85-90 decibels – average factory is 78 dB) leads to hearing loss. Noise pollution is the most common occupational hazard and can affect local ecology and natural environment.
OT03	Visual impact	Impact will vary depending on local surrounding and local community. Such impact can be considerable and often underestimated by those responsible for the site in question.
OT04	Other	Impact will vary depending on aspect and surrounding area affected.
AB	**Risk of abnormal activity**	
AB01	Risk of fire	Risk of explosions or fire from accidents or abnormal conditions. Could seriously affect WH & S and local environment.
AB02	Risk of spillage, leakage or uncontrolled discharge	Impact can affect local environment and water ecology, impact on groundwater quality and risk to WH & S.
AB03	Risk of spillage of special, hazardous or restricted substances	Discharge could affect human health, could contaminate land, ground water, local aquatic ecology and environment.
AB04	Risk to worker health and safety	Could seriously affect WH & S.
AB05	Other abnormalities	Impact will vary depending on abnormality and area affected.

Annex 4

ISO members

Albania (DSC)

Address:	Drejtoria e Standardizimit dhe Cilesise Rruga Mine Peza Nr. 143/3 Tirana
Telephone:	+ 355 42 2 62 55
Telefax:	+ 355 42 2 62 55
Telegram:	standardi tirana
Internet:	dsc@icc.al.eu.org

Algeria (INAPI)

Address:	Institut algérien de normalisation et de propriété industrielle 5, rue Abou Hamou Moussa B.P. 403 – Centre de tri Alger
Telephone:	+ 213 2 63 96 42
Telefax:	+ 213 2 61 09 71
Telex:	6 64 09 inapi dz
Telegram:	inapi-alger

Argentina (IRAM)

Address:	Instituto Argentino de Normalización Chile 1192 1098 Buenos Aires
Telephone:	+ 54 1 383 37 51
Telefax:	+ 54 1 383 84 63
Internet:	postmaster@iram.org.ar

Armenia (SARM)

Address:	Department for Standardization, Metrology and Certification Komitas Avenue 49/2 375051 Yerevan
Telephone:	+ 374 2 23 56 00
Telefax:	+ 374 2 28 56 20
Internet:	sarm@arminco.com

Australia (SAA)

Address:	Standards Australia 1 The Crescent Homebush – N.S.W. 2140
Postal address:	P.O. Box 1055 Strathfield – N.S.W. 2135
Telephone:	+ 61 2 9746 47 00
Telefax:	+ 61 2 9746 84 50
Telex:	2 65 14 astan aa
Internet:	intsect@saa.sa.telememo.au

Austria (ON)

Address:	Österreichisches Normungsinstitut Heinestrasse 38 Postfach 130 A-1021 Wien
Telephone:	+ 43 1 213 00
Telefax:	+ 43 1 213 00 650
Internet:	iro@tbxa.telecom.at

Bangladesh (BSTI)

Address: Bangladesh Standards and Testing
              Institution
              116/A, Tejgaon Industrial Area
              Dhaka – 1208
Telephone: + 880 2 88 14 62
Telefax: + 880 2 88 56 85
Telegram: besteye
Internet: bsti@bangla.net

Belarus (BELST)

Address: Committee for Standardization,
              Metrology and Certification
              Starovilensky Trakt 93
              Minsk 220053
Telephone: + 375 172 37 52 13
Telefax: + 375 172 37 25 88
Telex: 25 21 70 shkala
Internet: belst@mcsm.belpak.minsk.by

Belgium (IBN)

Address: Institut belge de normalisation
              Av. de la Brabançonne 29
              B-1000 Bruxelles
Telephone: + 32 2 738 01 11
Telefax: + 32 2 733 42 64
Internet: croon@ibn.be

Bosnia and Herzegovina (BASMP)

Address: Institute for Standardization,
              Metrology and Patents (BASMP)
              Dubrovacka 6
              CH-71000 Sarajevo
Telephone: 71 20 70 15
Telefax: 71 20 70 16

Brazil (ABNT)

Address: Associaçao Brasileira de Normas
              Técnicas
              Av. 13 de Maio, no 13, 28o andar
              20003-900 - Rio de Janeiro-RJ
Telephone: + 55 21 210 31 22
Telefax: + 55 21 532 21 43
Telex: 213 43 33 abnt br
Telegram: normatécnica rio
Internet: abnt@embratel.net.br

Bulgaria (BDS)

Address: Bulgarian Committee for
              Standardization and Metrology
              21, 6th September Str.
              1000 Sofia
Telephone: + 359 2 85 91
Telefax: + 359 2 80 14 02
Telex: 2 25 70 dks bg

Canada (SCC)

Address: Standards Council of Canada
              45 O'Connor Street, Suite 1200
              Ottawa, Ontario K1P 6N7
Telephone: + 1 613 238 32 22
Telefax: + 1 613 995 45 64
Internet: info@scc.ca

Chile (INN)

Address: Instituto Nacional de
              Normalización
              Matías Cousiño 64 - 6o piso
              Casilla 995 - Correo Central
              Santiago
Telephone: + 56 2 696 81 44
Telefax: + 56 2 696 08 74
Telegram: inn
Internet: inn@huelen.reuna.cl

China (CSBTS)

Address: China State Bureau of Technical
              Supervision
              4, Zhichun Road
              Haidian District
              P.O. Box 8010
              Beijing 100088
Telephone: + 86 10 6 203 24 24
Telefax: + 86 10 6 203 10 10
Telegram: 1918 beijing

Colombia (ICONTEC)

Address: Instituto Colombiano de Normas
              Técnicas y Certificación
              Carrera 37 52-95
              Edificio ICONTEC
              P.O. Box 14237
              Santafé de Bogotá
Telephone: + 57 1 315 03 77
Telefax: + 57 1 222 14 35
Telex: 4 25 00 icont co
Telegram: icontec
Internet: sicontec@col1.telecom.com.co

Costa Rica (INTECO)

Address:	Instituto de Normas Técnicas de Costa Rica
	Barrio González Flores
	Ciudad Científica
	San Pedro de Montes de Oca
	San José
Postal address:	P.O. Box 6189-1000
	San José
Telephone:	+ 506 283 45 22
Telefax:	+ 506 283 48 31
Internet:	inteco@sol.racsa.co.cr

Croatia (DZNM)

Address:	State Office for Standardization and Metrology
	Ulica grada Vukovara 78
	10000 Zagreb
Telephone:	+ 385 1 53 99 34
Telefax:	+ 385 1 53 65 98

Cuba (NC)

Address:	Oficina Nacional de Normalización
	Calle E No. 261 entre 11 y 13
	Vedado, La Habana 10400
Telephone:	+ 53 7 30 00 22
Telefax:	+ 53 7 33 80 48
Telex:	51 22 45 cen cu

Cyprus (CYS)

Address:	Cyprus Organization for Standards and Control of Quality
	Ministry of Commerce, Industry and
	Tourism
	Nicosia 1421
Telephone:	+ 357 2 37 50 53
Telefax:	+ 357 2 37 51 20
Telex:	22 83 mincomind cy
Telegram:	mincomind nicosia

Czech Republic (COSMT)

Address:	Czech Office for Standards, Metrology and Testing
	Biskupsky dvur 5
	110 02 Praha 1
Telephone:	+ 420 2 232 44 30
Telefax:	+ 420 2 232 43 73
Telex:	12 19 48 funm c
Telegram:	normalizace praha

Denmark (DS)

Address:	Dansk Standard
	Kollegievej 6
	DK-2920 Charlottenlund
Telephone:	+ 45 39 96 61 01
Telefax:	+ 45 39 96 61 02
Internet:	dansk.standard@ds.dk

Ecuador (INEN)

Address:	Instituto Ecuatoriano de Normalización
	Baquerizo Moreno 454 y
	Av. 6 de Diciembre
	Casilla 17-01-3999
	Quito
Telephone:	+ 593 2 56 56 26
Telefax:	+ 593 2 56 78 15
Internet:	inen1@inen.gov.ec

Egypt (EOS)

Address:	Egyptian Organization for Standardization and Quality Control
	2 Latin America Street
	Garden City
	Cairo
Telephone:	+ 20 2 354 97 20
Telefax:	+ 20 2 355 78 41
Telex:	9 32 96 eos un
Telegram:	tawhid
Internet:	moi@idsc.gov.eg

Ethiopia (EAS)

Address:	Ethiopian Authority for Standardization
	P.O. Box 2310
	Addis Ababa
Telephone:	+ 251 1 61 01 11
Telefax:	+ 251 1 61 31 77
Telex:	21725 ethsa eth
Telegram:	ethiostan

Finland (SFS)

Address:	Finnish Standards Association SFS
	P.O. Box 116
	FIN-00241 Helsinki
Telephone:	+ 358 9 149 93 31
Telefax:	+ 358 9 146 49 25
Internet:	sfs@sfs.fi

France (AFNOR)

Address:	Association française de normalisa-
	tion
	Tour Europe
	F-92049 Paris La Défense Cedex
Telephone:	+ 33 1 42 91 55 55
Telefax:	+ 33 1 42 91 56 56
Telex:	61 19 74 afnor f
Telegram:	afnor courbevoie

Germany (DIN)

Address:	DIN Deutsches Institut für
	Normung
	Burggrafenstrasse 6
	D-10787 Berlin
Postal address:	D-1072 Berlin
Telephone:	+ 49 30 26 01-0
Telefax:	+ 49 30 26 01 12 31
Telex:	18 42 73 din d
Telegram:	deutschnormen berlin
Internet:	postmaster@din.de

Ghana (GSB)

Address:	Ghana Standards Board
	P.O. Box M 245
	ACCRA
Telephone:	+ 233 21 50 00 65
Telefax:	+ 233 21 50 00 92

Greece (ELOT)

Address:	Hellenic Organization for
	Standardization
	313, Acharnon Street
	GR-111 45 Athens
Telephone:	+ 30 1 228 00 01
Telefax:	+ 30 1 228 30 34
Telex:	21 96 21 elot gr
Telegram:	elotyp-athens
Internet:	elotinfo@elot.gr

Hungary (MSZT)

Address:	Magyar Szabványügyi Testület
	Üllöi út 25
	Pf. 24.
	H-1450 Budapest 9
Telephone:	+ 36 1 218 30 11
Telefax:	+ 36 1 218 51 25
Internet:	sze1545@helka.iif.hu

Iceland (STRI)

Address:	Icelandic Council for
	Standardization
	Keldnaholt
	IS-112 Reykjavik
Telephone:	+ 354 570 71 50
Telefax:	+ 354 570 71 11
Internet:	stri@stri.is

India (BIS)

Address:	Bureau of Indian Standards
	Manak Bhavan
	9 Bahadur Shah Zafar Marg
	New Delhi 110002
Telephone:	+ 91 11 323 79 91
Telefax:	+ 91 11 323 40 62
Telex:	316 58 70 bis in
Telegram:	manaksanstha
Internet:	bisind@del2.vsnl.net.in

Indonesia (DSN)

Address:	Dewan Standardisasi Nasional -
	DSN
	(Standardization Council of
	Indonesia)
	c/o Pusat Standardisasi - LIPI
	Jalan Jend. Gatot Subroto 10
	Jakarta 12710
Telephone:	+ 62 21 522 16 86
Telefax:	+ 62 21 520 65 74
Telex:	6 28 75 pdii ia
Telegram:	lipi jakarta
Internet:	pustan@rad.net.id

Iran, Islamic Republic of (ISIRI)

Address:	Institute of Standards and
	Industrial Research of Iran
	P.O. Box 31585-163
	Karaj
Telephone:	+ 98 261 22 60 31-5
Telefax:	+ 98 261 22 50 15
Telex:	21 54 42 stan ir
Telegram:	standinst

Ireland (NSAI)

Address:	National Standards Authority of Ireland
	Glasnevin
	Dublin-9
Telephone:	+ 353 1 807 38 00
Telefax:	+ 353 1 807 38 38
Telegram:	research, dublin
Internet:	nsai@nsai.ie

Israel (SII)

Address:	Standards Institution of Israel
	42 Chaim Levanon Street
	Tel Aviv 69977
Telephone:	+ 972 3 646 51 54
Telefax:	+ 972 3 641 96 83
Internet:	standard@netvision.net.il

Italy (UNI)

Address:	Ente Nazionale Italiano di Unificazione
	Via Battistotti Sassi 11/b
	I-20133 Milano
Telephone:	+ 39 2 70 02 41
Telefax:	+ 39 2 70 10 61 06
Telegram:	unificazione
Internet:	webmaster@uni.unicei.it

Jamaica (JBS)

Address:	Jamaica Bureau of Standards
	6 Winchester Road
	P.O. Box 113
	Kingston 10
Telephone:	+ 1 809 926 31 40-6
Telefax:	+ 1 809 929 47 36
Telex:	22 91 stanbur ja
Telegram:	stanbureau
Internet:	jbs@toj.com

Japan (JISC)

Address:	Japanese Industrial Standards Committee
	c/o Standards Department
	Ministry of International Trade and Industry
	1-3-1, Kasumigaseki, Chiyoda-ku
	Tokyo 100
Telephone:	+ 81 3 35 01 20 96
Telefax:	+ 81 3 35 80 86 37

Kenya (KEBS)

Address:	Kenya Bureau of Standards
	Off Mombasa Road
	Behind Belle Vue Cinema
	P.O. Box 54974
	Nairobi
Telephone:	+ 254 2 50 22 10/19
Telefax:	+ 254 2 50 32 93
Telex:	2 52 52 viwango
Telegram:	kenstand
Internet:	kebs@arso.gn.apc.org

Korea, Dem. P. Rep. of (CSK)

Address:	Committee for Standardization of the Democratic People's Republic of Korea
	Zung Gu Yok Seungli-Street
	Pyongyang
Telephone:	+ 85 02 57 15 76
Telex:	59 72 tech kp
Telegram:	standard

Korea, Republic of (KNITQ)

Address:	Korean National Institute of Technology and Quality
	1599 Kwanyang-dong
	Dongan-ku, Anyang-city
	Kyonggi-do 430-060
Telephone:	+ 82 3 43 84 18 61
Telefax:	+ 82 3 43 84 60 77

Libyan Arab Jamahiriya (LNCSM)

Address:	Libyan National Centre for Standardization and Metrology
	Industrial Research
	Centre Building
	P.O. Box 5178
	Tripoli
Telephone:	+ 218 21 369 30 74
Telefax:	+ 218 21 369 30 71
Telex:	2 05 49 ncsm

Malaysia (DSM)

Address:	Department of Standards Malaysia
	21st Floor, Wisma MPSA
	Persiaran Perbandaran
	40675 Shah Alam
	Selangor Darul Ehsan
Telephone:	+ 60 3 559 80 33
Telefax:	+ 60 3 559 24 97
Internet:	central@dsm4.gov.my

Mauritius (MSB)

Address: Mauritius Standards Bureau
             Moka
Telephone: + 230 433 36 48
Telefax: + 230 433 51 50

Mexico (DGN)

Address: Dirección General de Normas
             Calle Puente de Tecamachalco No
             6
             Lomas de Tecamachalco
             Sección Fuentes
             Naucalpan de Juárez
             53 950 Mexico
Telephone: + 52 5 729 93 00
Telefax: + 52 5 729 94 84
Telex: 177 58 40 imceme
Telegram: secofi/147
Internet: cidgn@secofi.gob.mx

Mongolia (MNCSM)

Address: Mongolian National Centre for
             Standardization and Metrology
             Peace street 46A
             Ulaanbaatar-51
Telephone: + 976 1 35 83 49
Telefax: + 976 1 35 80 32
Internet: MNCSM@magicnet.mn

Morocco (SNIMA)

Address: Service de normalisation indus-
             trielle marocaine
             Ministère du commerce, de l'indus-
             trie et l'artisanat
             Quartier administratif
             Rabat Chellah
Telephone: + 212 7 76 37 33
Telefax: + 212 7 76 62 96
Telex: 36 872

Netherlands (NNI)

Address: Nederlands Normalisatie-instituut
             Kalfjeslaan 2
             P.O. Box 5059
             NL-2600 GB Delft
Telephone: + 31 15 2 69 03 90
Telefax: + 31 15 2 69 01 90
Telex: 3 81 44 nni nl
Telegram: normalisatie delft
Internet: info@nni.nl

New Zealand (SNZ)

Address: Standards New Zealand
             Standards House
             155 The Terrace
             Wellington 6001
Postal address: Private Bag 2439
             Wellington 6020
Telephone: + 64 4 498 59 90
Telefax: + 64 4 498 59 94
Internet: snz@standards.synet.net.nz

Nigeria (SON)

Address: Standards Organisation of Nigeria
             Federal Secretariat
             Phase 1, 9th Floor
             Ikoyi
             Lagos
Telephone: + 234 1 68 26 15
Telefax: + 234 1 68 18 20

Norway (NSF)

Address: Norges Standardiseringsforbund
             Drammensveien 145 A
             Postboks 353 Skoyen
             N-0212 Oslo
Telephone: + 47 22 04 92 00
Telefax: + 47 22 04 92 11
Internet: firmapost@nsf.telemax.no telemax.
             no

Pakistan (PSI)

Address: Pakistan Standards Institution
             39 Garden Road, Saddar
             Karachi-74400
Telephone: + 92 21 772 95 27
Telefax: + 92 21 772 81 24
Telegram: peyasai

Panama (COPANIT)

Address: Comisión Panameña de Normas
             Industriales y Técnicas
             Ministerio de Comercio e
             Industrias
             Apartado Postal 9658
             Panama, Zona 4
Telephone: + 507 2 27 47 49
Telefax: + 507 2 25 78 53

Philippines (BPS)

Address: Bureau of Product Standards
Department of Trade and Industry
361 Sen. Gil J. Puyat Avenue
Makati
Metro Manila 1200
Telephone: + 63 2 890 49 65
Telefax: + 63 2 890 49 26
Internet: dtibpsrp@sequel.net

Poland (PKN)

Address: Polish Committee for
Standardization
ul. Elektoralna 2
P.O. Box 411
PL-00-950 Warszawa
Telephone: + 48 22 620 54 34
Telefax: + 48 22 620 54 34

Portugal (IPQ)

Address: Instituto Português da Qualidade
Rua C à Avenida dos Três Vales
P-2825 Monte de Caparica
Telephone: + 351 1 294 81 00
Telefax: + 351 1 294 81 01
Internet: ipqmail@ipqm.ipqgtw-ms.mail-
pac.pt

Romania (IRS)

Address: Institutul Român de Standardizare
Str. Jean-Louis Calderon Nr. 13
Cod 70201
R-Bucuresti 2
Telephone: + 40 1 211 32 96
Telefax: + 40 1 210 08 33

Russian Federation (GOST R)

Address: Committee of the Russian
Federation for Standardization,
Metrology and Certification
Leninsky Prospekt 9
Moskva 117049
Telephone: + 7 095 236 40 44
Telefax: + 7 095 237 60 32
Telex: 41 13 78 gost su
Telegram: moskva standart
Internet: gosstandart@sovcust.sprint.com

Saudi Arabia (SASO)

Address: Saudi Arabian Standards
Organization
Imam Saud Bin Abdul Aziz Bin
Mohammed
Road (West End)
P.O. Box 3437
Riyadh 11471
Telephone: + 966 1 452 00 00
Telefax: + 966 1 452 00 86
Telex: 40 16 10 saso sj
Telegram: giasy

Singapore (PSB)

Address: Singapore Productivity and
Standards Board (PSB)
1 Science Park Drive
Singapore 118221
Telephone: + 65 278 66 66
Telefax: + 65 776 12 80

Slovakia (UNMS)

Address: Slovak Office of Standards,
Metrology and Testing
Stefanovicova 3
814 39 Bratislava
Telephone: + 42 17 39 10 85
Telefax: + 42 17 39 10 50

Slovenia (SMIS)

Address: Standards and Metrology Institute
Ministry of Science and
Technology
Kotnikova 6
SI-1000 Ljubljana
Telephone: + 386 61 178 30 00
Telefax: + 386 61 178 31 96
Internet: smis@usm.mzt.si

South Africa (SABS)

Address: South African Bureau of Standards
1 Dr Lategan Rd, Groenkloof
Private Bag X191
Pretoria 0001
Telephone: + 27 12 428 79 11
Telefax: + 27 12 344 15 68
Telex: 32 13 08 sa
Telegram: comparator
Internet: postmaster@sabs.co.za

Spain (AENOR)

Address: Asociación Española de
Normalización y Certificación
Génova, 6
E-28004 Madrid
Telephone: + 34 1 432 60 00
Telefax: + 34 1 310 49 76
Telegram: aenor

Sri Lanka (SLSI)

Address: Sri Lanka Standards Institution
53 Dharmapala Mawatha
P.O. Box 17
Colombo 3
Telephone: + 94 1 32 60 51
Telefax: + 94 1 44 60 18
Telegram: pramika
Internet: dg@hoslsi.ac.lk

Sweden (SIS)

Address: SIS - Standardiseringen i Sverige
St Eriksgatan 115
Box 6455
S-113 82 Stockholm
Telephone: + 46 8 610 30 00
Telefax: + 46 8 30 77 57
Internet: info@sis.se

Switzerland (SNV)

Address: Swiss Association for
Standardization
Mühlebachstrasse 54
CH-8008 Zurich
Telephone: + 41 1 254 54 54
Telefax: + 41 1 254 54 74
Telegram: normbureau
Internet: post@snv.snv.inet.ch

Syrian Arab Republic (SASMO)

Address: Syrian Arab Organization for
Standardization and Metrology
P.O. Box 11836
Damascus
Telephone: + 963 11 445 05 38
Telefax: + 963 11 512 82 14
Telex: 41 19 99 sasmo
Telegram: systand

Tanzania, United Rep. of (TBS)

Address: Tanzania Bureau of Standards
Ubungo Area
Morogoro Road/Sam Nujoma
Road
Dar es Salaam
Postal address: P.O. Box 9524
Dar es Salaam
Telephone: + 255 51 4 32 98
Telefax: + 255 51 4 32 98
Telex: 4 16 67 tbs tz
Telegram: standards
Internet: tbs@costech.gn.apc.org

Thailand (TISI)

Address: Thai Industrial Standards Institute
Ministry of Industry
Rama VI Street
Bangkok 10400
Telephone: + 66 2 245 78 02
Telefax: + 66 2 247 87 41
Telex: 8 43 75 minidus th (attention tisi)
Telegram: thastan
Internet: thaistan@tisi.go.th

The former Yugoslav Republic of Macedonia (ZSM)

Address: Zavod za standardizacija i
metrologija (ZSM)
Ministry of Economy
Samoilova 10
91000 Skopje
Telephone: + 389 91 13 11 02
Telefax: + 389 91 11 02 63

Trinidad and Tobago (TTBS)

Address: Trinidad and Tobago Bureau of
Standards
#2 Century Drive
Trincity Industrial Estate
Tunapuna
Postal address: P.O. Box 467
Port of Spain
Telephone: + 1 868 662 88 27
Telefax: + 1 868 663 43 35
Telegram: qualassure
Internet: ttbs@opus-networx.com

Tunisia (INNORPI)

Address: Institut national de la normalisation
               et de la propriété industrielle
               B.P. 23
               1012 Tunis-Belvédère
Telephone: + 216 1 78 59 22
Telefax: + 216 1 78 15 63

Turkey (TSE)

Address: Türk Standardlari Enstitüsü
               Necatibey Cad. 112
               Bakanliklar
               TR-06100 Ankara
Telephone: + 90 312 417 83 30
Telefax: + 90 312 425 43 99
Telex: 4 20 47 tse-tr
Telegram: standard
Internet: didb@tse.org.tr

USA (ANSI)

Address: American National Standards
               Institute
               11 West 42nd Street
               13th floor
               New York, N.Y. 10036
Telephone: + 1 212 642 49 00
Telefax: + 1 212 398 00 23
Internet: info@ansi.org

Ukraine (DSTU)

Address: State Committee of Ukraine for
               Standardization, Metrology and
               Certification
               174 Gorkiy Street
               GSP, Kyiv-6, 252650
Telephone: + 380 44 226 29 71
Telefax: + 380 44 226 29 70

United Kingdom (BSI)

Address: British Standards Institution
               389 Chiswick High Road
               GB-London W4 4AL
Telephone: + 44 181 996 90 00
Telefax: + 44 181 996 74 00
Internet: info@bsi.org.uk

Uruguay (UNIT)

Address: Instituto Uruguayo de Normas
               Técnicas
               San José 1031 P.7
               Galeria Elysée
               Montevideo
Telephone: + 598 2 91 20 48
Telefax: + 598 2 92 16 81
Telex: 2 31 68 ancap uy
Internet: unit@adinet.com.uy

Uzbekistan (UZGOST)

Address: Uzbek State Centre for
               Standardization, Metrology and
               Certification
               Ulitsa Farobi, 333-A
               700049 Tachkent
Telephone: + 7 371 2 46 17 10
Telefax: + 7 371 2 46 17 11
Telex: 11 63 82 fasad

Venezuela (COVENIN)

Address: Comisión Venezolana de Normas
               Industriales
               Avda. Andrés Bello-Edf. Torre
               Fondo
               Común
               Piso 12
               Caracas 1050
Telephone: + 58 2 575 22 98
Telefax: + 58 2 574 13 12
Telex: 2 42 35 minfo vc
Telegram: covenindus
Internet: covenin@dino.conicit.ve

Viet Nam (TCVN)

Address: Directorate for Standards and
               Quality
               70, Tran Hung Dao Street
               Hanoi
Telephone: + 84 4 826 62 20
Telefax: + 84 4 826 74 18
Internet: thien@netnam.org.vn

Yugoslavia (SZS)

Address: Savezni zavod za standardizaciju
            Kneza Milosa 20
            Post Pregr. 933
            YU-11000 Beograd
Telephone: + 381 11 64 35 57
Telefax: + 381 11 68 23 82
Telegram: standardizacija
Internet: nauka@hera.smrnzs.sv.gov.yu

Zimbabwe (SAZ)

Address: Standards Association of
            Zimbabwe
            P.O. Box 2259
            Harare
Telephone: + 263 4 88 20 17
Telefax: + 263 4 88 20 20
Telegram: saca

Abbreviations and acronyms

BS	British Standard
BSI	British Standards Institute
BATNEEC	Best available technique not entailing excessive cost
BOD	Biological oxygen demand
CERES	Coalition for Environmental Responsible Economies
CFC	Chloroflourocarbon
CO$_2$	Carbon dioxide
dB	Decibel
DFE	Design for the environment
DIS	Draft international standard
DGP	Dark grains plant
DTI	UK Department of Trade and Industry
EC	European Commission
EI	Entropy International
EIA	Environmental impact assessment
EMAS	Eco-management and audit scheme
EMM	Environmental management manual
EMS	Environmental management system
EU	European Union
EVABAT	Economically viable application of best available technology
GATT	General Agreement on Tariffs and Trade
ICC	International Chamber of Commerce
IER	Initial environmental review
ISO	International Organization for Standardization

LCA	Life cycle analysis (sometime referred to as life cycle assessment)
MVR	Mechanical vapour recompressor
NO$_x$	Oxides of nitrogen (N_2O, NO and NO_2)
OJ	Official Journal
PCB	Polychlorinated biphenyls
PVC	Polyvinyl chloride
QMS	Quality management system
SAGE	Strategic Advisory Group on the Environment
SEPA	Scottish Environmental Protection Agency
SO$_x$	Oxides of sulphur (SO_2 and SO_3)
TAG	Technical advisory group
TC	Technical committee
UD	United Distillers
UK	United Kingdom
UMGD	United Malt and Grain Distillers
UN	United Nations
UNEP	United Nations Environment Programme
US EPA	United States Environmental Protection Agency
UV	Ultra violet
WESP	Wet electrostatic precipitator
WHS	Worker health and safety

Glossary

Acidification: Unusually acidic terrestrial and aquatic ecosystems as a result of acid rain. Acid rain, pH 3.0 - 5.5, is caused by pollution from the combustion of fossil fuels, and the resulting accumulations of oxides of nitrogen and sulphur in the atmosphere.

Aspect: *See* Environmental aspect.

Correction. The act of developing or improving where nonconformance has been identified.

Document control. The set of procedures by which you ensure your EMS documents are organized, current, locatable and 'controlled' in a manner that guarantees their efficacy.

Environmental aspect: Activities, products or processes that interact, or could interact, with the environment, either directly or indirectly, to cause an environmental impact. Aspects are things like water use, flue gas emission from a furnace, energy use or effluent discharge.

Environmental audit (EMS Audit). The process of assessing whether or not an environmental management system (EMS) is functioning as it should by comparing it against predetermined criteria and assessing conformance to those criteria.

Environmental communication. Environmental communication falls into two categories – internal communication and external communication. Internal communication is the communication between the various levels and functions involved in the development, implementation and maintenance of your EMS. Internal environmental communication would include things such as training personnel on the environmental policy, interaction between personnel with identified responsibilities for maintaining the EMS, informing upper management on changes to or results of your EMS. External communication is essentially communication with those who are affected by your environmental aspects and/or your EMS. Your environmental policy is also a form of external communication.

Environmental impact: Changes, both positive and negative, to the environment as a result of an environmental aspect(s). Impacts are things like a change in mean stream temperature as a result of effluent discharge to that stream, atmospheric acidification as a result of flue gas emissions, or the contamination or degradation of land as a result of waste disposal.

Environmental impact assessment (EIA). The process of assessing the environmental impacts (effects) of an activity (often in the future), such as developing a new production process or building a dam.

Environmental management. The process of minimizing the environmental impacts of your organization by controlling the aspects of your activities, products or processes that cause, or could cause, those impacts to the environment.

Environmental management manual. The central tool or reference for key documents that are required for maintaining and auditing your EMS over time.

Environmental management programme. A documented plan that identifies how objectives and targets will be met, who is responsible for each of the activities required to meet those objectives and targets and when those activities will be completed.

Environmental management system (EMS). The organizational structures, activities, roles, responsibilities, procedures and resources that together enable an organization to minimize their environmental identified significant impacts by controlling the environmental aspects that cause, or could cause, the impacts identified.

Environmental objectives. The broad goals that your organization sets in order to improve environmental performance. Environmental objectives are goals such as 'reduce water use' or 'improve energy efficiency'.

Environmental policy. A formal and documented set of principles and intentions with respect to the environment. The environmental policy is the guiding document for corporate environmental improvement.

Environmental targets. Set performance measurements that must be met to realize a given objective. Targets are measurable and quantifiable statements such as 'By 10 cubic meters/day' or '50% by the year 2000'.

Eutrophication: The process of depleting a body of freshwater of oxygen and thus starving (and killing off) the higher life forms it contains (fish, frogs etc.) Eutrophication is caused by a rapid growth of primary organic matter (such as algae and plankton) and the corresponding acceleration of oxygen demanded by them. Eutrophication results from an abundant supply of plant nutrients (mostly phosphates and nitrates from fertilizer run-off, sewage treatment discharged, etc.)

Impact: *See* Environmental impact.

Initial environmental review (IER). A systematic identification and documentation of the significant environmental impacts (or potential for impacts) associated directly or indirectly with your organization's activities, products and processes.

ISO 14000: A series of international standards for environmental management. It is the first such series that allows organizations from around the world to pursue environmental efforts and measure perfomance according to internationally accepted criteria.

ISO 14001: The first in the 14000 series. It specifies the requirements of an environmental management system. ISO 14001 is a voluntary standard and was developed by the International Organization for Standardization (ISO). ISO 14001 is intended to be applicable to 'all types and sizes of organizations and to accommodate diverse geographical, cultural and social conditions'. The overall aim of both ISO 14001 and the other standards in the 14000 series is to support environmental protection and the prevention of pollution in harmony with socio-economic needs.

Life cycle analysis. The assessment of the environmental impact of a product (or service) during its full life. This would include an assessment of the impact of raw material extraction, product design and development, production, product use and final product disposal.

Management review. The formal evaluation, by management, of the audit findings and the degree to which organization's environmental policy, objectives and targets, and procedures are functioning as tools to improve environmental performance.

Monitoring and measuring. The means by which an organization identifies its progress toward the minimization of the environmental impact of its activities, products and processes.

Nonconformance. The situation where essential components of your EMS are absent or dysfunctional or where there is insufficient control of your activities, products or processes to the extent that these absences compromise your policy, objectives and targets, management programmes and the functionality of your EMS.

Operational control. The set of procedures that ensure your operations (aspects) are controlled. Operational control improves environmental performance by controlling aspects and minimizing correlating environmental impacts caused by those operations.

Ozone depletion: Depletion of the naturally occurring layer of stratospheric ozone that protects the earth from harmful ultraviolet radiation from the sun. Ozone depletion results from emission to the atmosphere of ozone-depleting substances (ODS) such as compounds that contain chlorine and bromine.

Prevention. The act of ensuring that nonconformances will not re-occur.

Procedures. The step-by-step instructions that, if carried out properly, will control both your environmental management system and your organization's activities, products and processes (aspects). This control will in turn minimize correlating environmental impacts and thus improve the overall environmental performance of your organization.

Records. The essential documents that constitute a functional EMS. Records include the environmental policy, objectives and targets, environmental management programmes, procedures and the documents that contain data by which you benchmark the performance of your EMS.

Register of aspects and impacts. The register of environmental aspects and impacts is the documented record of significant environmental aspects and correlating impacts that the organization must control and minimize to improve its overall corporate environmental performance. The environmental aspects and impacts contained in the register should be those that are identified as being significant in your initial environmental review.

Register of legislation and regulations. The register of environmental legislation and regulations is a list of all the relevant environmental legislation and regulations by which your organization is bound and with which your identified significant environmental aspects and impacts are associated.

Responsibility. Responsibility refers to the roles, authorities and interrelations of the key personnel required to ensure the efficacy of an environmental management system and its chosen structure.

Solvent: A general term for any liquid that is used to dissolve other substances. Solvents such as organochlorines, commonly used as cleaning agents, are responsible for a considerable amount of groundwater contamination. The most common industrial solvents include acetic acid, acetone, benzene, cyclohexanol, ethanol, furfural, glycerol, hexane, isopropanol, methanol, methylethylketone (MEK), n-propanol, toluene and trichlorethylene.

Stakeholders. People or organizations with an interest in the environmental impact of your organization's activities, products and services. This can include government regulators and inspectors, investors, bankers, shareholders, insurers, employees, the local community, customers, clients, consumers, non-governmental organizations, environmental groups and the public.

Structure. The administrative form of an environmental management system.

Verification. Action by a third party (EMAS verifier), demonstrating that adequate confidence is provided that a duly identified EMS is in compliance with the respective requirements of a specific standard or other normative document (Council Regulation (EEC) No 1836/93).

Volatile organic compounds (VOC): Compounds such as solvents, propylene, acetone, styrene, benzene and ethylene which evaporate and contribute to air pollution and photochemical smogs.

Index

A

Abbreviations and acronyms, 397
Accidents (preventing), 206
Accidents and incidents, 73
Aspects and impacts
  register, 95
  review, 61
Audit findings, 235
Audit follow-up, 232
Audit plan, 232
Audit report, 232
Audit trails, 235
Audits, 231
Authorizations (review), 77
Awareness, 167

B

Benefits of EMS, 12
BS 5750, 16
BS 7750, 16

C

Case study, 269
Certification of EMS, 17
Certification process, 18
Check-lists for IER, 333
Check-lists
  Document control, 197
  Emergency preparedness and response, 209
  EMS audit, 256
  Environmental communication, 180

Check-lists
  Environmental legislation and regulations, 128
  Environmental management manual, 191
  Environmental management programme, 148
  Environmental objectives and targets, 138
  Environmental policy, 112
  Environmental procedures, 165
  Environmental records, 229
  Environmental structure and responsibilities, 156
  Environmental training and awareness, 173
  Initial environmental review, 93
  Management review, 265
  Monitoring and measuring, 215
  Nonconformance, correction and prevention, 223
  Operational control, 203
  Register of environmental aspects and impacts, 104
Closing meeting, 236
Codes of practice (review), 77
Collecting evidence, 234
Communication, 175
Company being reviewed, 40
Conducting an EMS audit, 234
Conducting the IER, 32
Continual improvement, 3
Controlled documents, 194
Controlled status, 194
Correction, 217

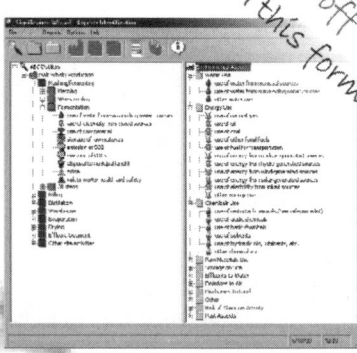
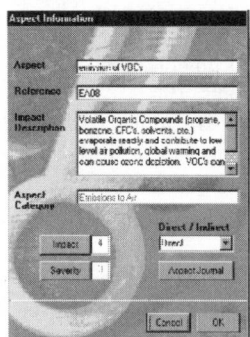
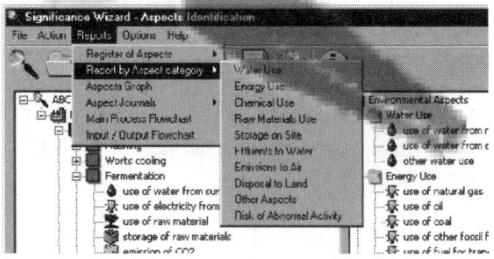